I0058234

Humanity's Future

How Technology Will Change Us

by

Jay Friedenberg

© 2014 Jay Friedenberg

All Rights Reserved

No part of this book may be reproduced or transmitted in any form or by any means, graphic, electronic, or mechanical, including photocopying, recording, taping, or by any information storage retrieval system, without the written permission of the publisher—except by a reviewer who may quote brief passages for magazines, newspapers, and internet sites. The cover of this book may be shown on the internet or in promoting the book elsewhere.

Jay Friedenberg
Professor of Psychology
Manhattan College
Department of Psychology
jay.friedenberg@manhattan.edu
917-882-5185

TABLE OF CONTENTS

INTRODUCTION. BRAVE NEW WORLDS

This book is a collection of essays written on a variety of topics relevant to the future of humankind. The chapters are organized foundationally starting with how it is we can know and understand reality. This is followed by descriptions of future technologies and the new science that will drive them. Then we take a look at ourselves, how smart we are and how intelligent our machines may become. The final sections paint a larger picture, examining civilization with a focus on what we've done wrong historically and how we may correct such mistakes. The essays are self-contained so the reader can skip around and read them in a different order should they desire.

Each essay starts by defining the topics at hand along with a brief description of the current state of the art. This is followed by predictions as to what the future might hold in this area and how that outcome might affect us psychologically as individuals or collectively as a society. The farther ahead one gets the more difficult it is to predict the future and in many cases I attempt to only do short range forecasting based on trend analysis. My concern is not so much on getting these events right but on what our reactions to them might be. These essays are a mixture. They contain selected summaries of the literature on a topic as well as my own opinions on them. Consider them factually inspired editorials.

I have kept the essays deliberately short to make them easier to digest. Most of them are about two pages in length. My approach is one of breadth and not depth. In some cases there is overlap and a single topic may be addressed in more than one essay. This is certainly the case with technology, which is a theme that runs throughout the book. In some instances a single topic like intelligence may be broken down into separate sections. For example there are several essays on intelligence, one on human intelligence and others on computer intelligence.

The perspective I adopt here is pro-reason, pro-science and pro-technology. I am a big believer that we humans can get at the truth and the good. We have and will continue to understand the world through science and manipulate it using applied engineering and technology. Our successes in this regard are impressive. It is amazing how much we have accomplished in just the past century. Cell phones and the World Wide Web would seem like science fiction to somebody living in even a time as recent as the 19th century. As Arthur C. Clarke said: "Any sufficiently advanced technology is indistinguishable from magic".

Beyond this though, we are now reaching the point where we can understand and change ourselves to the same degree as the environment. Using genetic engineering and pharmacology we can eliminate disease and unwanted traits while making ourselves healthier and more intelligent. We are only a few years away from being able to simulate computationally large portions of the human brain and potentially of creating an artificial intelligence that can rival and exceed our own.

All this may not come without a cost and is fraught with questions. Will we feel useless in a future where most tasks are automated and robots do all the work? Will society experience alienation and angst, potentially collapsing into a state of hedonism, decadence and corruption? How do we improve ourselves? Who gets to decide? How can we advance civilization and correct for the many mistakes we have made in the past like genocide and war? Can we create a utopia? I certainly don't pretend to have answers to all of these questions but we can explore some of the issues surrounding them.

This book is intended for the lay reader and does not assume background knowledge in any of the specialized fields. Terms are defined and the descriptions are not technical. To preserve readability there are no references within the body of the text other than the mention of a seminal work. At the end are a list of suggested books and web links for those who want to explore more. For those who like to take their learning with a dose of fiction, I have also included a list of recommended science fiction media based on selected topics.

CHAPTER 1. KNOWLEDGE AND KNOWING

REALITY AND SIMULATION

Is reality real? Are you in a computer simulation? Do you have a physical body and are experiencing reality or are you being fed information that makes it seem like you are in reality? These questions were the crux of *The Matrix* movie trilogy in which a future world of machines has enslaved humanity and simulates reality for them while using their bodies to generate energy. A frightening possibility but could it be true? Surprisingly the answer is yes.

Philosophers call this the "brain in a vat" scenario. Imagine that aliens have removed your brain and it is floating in a vat of fluid. Attached to it are inputs that feed from a powerful computer into the sensory parts of your brain. One cable feeds your visual system images. Another provides your auditory system with sounds. The computer generates an artificial set of stimuli such that you see and hear a world that isn't there. Conversely, there are output cables from the motor parts of your brain that send information to your muscles. If your brain issues a command to move your legs to walk the computer would pick up those commands and change the virtual display to change with your intended motions.

Now of course most people scoff at this idea and rightly so. According to the principle of Occam's razor, the simplest explanation is usually the correct one. To postulate that there are aliens doing this to us complicates an explanation of our consciousness so it is probably wrong. However as we can only know things through what we experience we cannot "get outside" our experience to prove it is false. So this philosophical thought experiment will forever remain untested.

Nick Bostrom of the Future of Humanity Institute proposes a similar scenario. Instead of a brain in a vat he argues that we could exist as a disembodied program inside an alien computer system. Therefore we would not only be without bodies but we would also be without brains. Our experience would then correspond to the execution of the program in the computer. Bostrom argues that this scenario is entirely possible by outlining three propositions. The first is that a significant fraction of all species eventually become technologically mature. The second is that some portion of those species will use their computational resources to run computer simulations of minds like ours. The third statement is that we are in such a simulation. If the first two statements are true then the third can also be true.

Why would aliens ever decide to do this? Bostrom says they might want to do it for artistic, scientific or recreational reasons. Just as we like to watch video games or movies, they might enjoy experiencing these kinds of simulations. If at some point in the simulation they decided they wanted to let us know, they could do so. Perhaps a window could open up in front of us with a voice or visual text announcing the fact!

If we were indeed inside a world simulation and were informed about it, how might this affect our behavior? There are a couple of options here. First of all we might go crazy. Schizophrenics have these sorts of delusions and it in part explains their paranoia. Another possibility is that we could simply accept the situation and get on with our "lives". If the rules of the simulation did not change then we could simply continue acting in the same manner we always have. Only in a situation where the rules governing the simulation changed, i.e., the revocation of a principle such as gravity, would we need to be concerned.

PROOF AND UNDERSTANDING

Proof and truth apply to statements that describe the world. "The moon is made of cheese" is an example. If a statement can be tested, we say it falls into the category of a hypothesis. Let's say we send a man to the moon who comes back with a soil sample and the sample is not made of cheese. In this case we have under the conditions falsified the statement or at the very least failed to confirm it. There are also statements that may never be testable. An example of this is "God exists". Technically this is called a non-testable assertion. We cannot test it because God as conceptualized in most religions cannot be measured. There is no experiment that could be designed to verify or falsify this statement because there is no evidence for us to go on.

So we cannot prove things for which there is no evidence. If there is no evidence for something we must acknowledge that it might not exist. However, there is another saying that goes: "Absence of evidence is not evidence of absence". This means that one can make a reasoned or probabilistic argument for the existence of certain things such as extraterrestrial life. As discussed elsewhere in this book we can take the knowledge we have of the universe and use it to make an informed guess for the existence of aliens. We cannot do the same for the existence of things like alternate universes that by their nature cannot be tested.

Only recently in human history do we have methods that allow us access to truth. These include reason, logic and the scientific method. Prior to this there were many routes to knowledge, not all resulting in accurate conclusions. Mysticism is one of these routes. Many prophets claim to have arrived at the truth by having experienced a "vision". This so called vision in many cases being a communication from a God or spirit. Another one of these routes is intuition, an ill-specified emotional process by which some thought just "feels" right.

We are spiritual and emotional creatures in addition to being cognitive so it makes sense that we might accept these as a means towards knowledge. Many people today however contradict themselves because they simultaneously hold verified and unverified beliefs. Quite a few people accept the tenets of their religion and will admit to believing in heaven, hell, angels, devils and souls, etc. while at the same time accepting the results of scientific research. It is as if they compartmentalize themselves with one part accepting religious doctrine on faith and another part accepting modern methods of knowing.

It is rare to find a person who is consistently rational and logical, who believes only in the evidence of their senses, what they can glean from their own deductive reasoning processes and what they read from valid sources. This "Spock-like" character as portrayed in the *Star Trek* series doesn't seem to be human. He or she may be right most of the time, but they seem to do so at the expense of their emotions, repressing and controlling them. Frequently in these stories they are intensely disliked for having turned themselves into "human computers", in the process somehow losing their soul or what makes them human.

It is possible to be both consistently rational and appropriately emotional but it requires self-control. In the future we should strive to be more like this. There are particular implications here for education. Students at all levels must be taught how to think critically and analytically. They should also be schooled in appropriate forms of evidence and how to interpret them. For example, when a correlation between two variables (like eating carrots and living longer) is presented on the evening news they ought to know that means there is a relationship between the two variables but it may not be a causal one.

A great way to learn how to reason effectively is to take math and computer programming courses. Both of these are examples of formal systems. Formal systems contain a set number of representative symbols that can be combined using operators under specified rules. Courses on critical thinking are also valuable as they teach students how to check assumptions, understand valid forms of evidence and determine inferences, implications and consequences of an argument. There are textbooks on critical thinking that describe fallacies and biases people make when reasoning. Problem solving of all types should also be encouraged in the educational system. Schools of the future should teach us not just facts, but how to acquire them.

Let's round off this essay by discussing thought experiments. These are hypothetical situations often devised by philosophers to try to tease out an answer to a problem. One classic example is that a train engineer sees an out of control train coming down a track that will hit a group of people. The engineer can either pull a lever to switch the train to another track killing his wife or do nothing and let it hit the group. What should he do? Thought experiments like this are contrived and don't occur in the real world but they do help us to think about an issue and how people might react under those circumstances. Philosophy in general is better at generating hypotheses and asking questions than it is at providing answers. It is best utilized in conjunction with empirical testing.

EDUCATION

What will the future of education be like? Will software and robots take over, making flesh-and-blood teachers a thing of the past? Not entirely. One scenario would have multi-media play an important role. One could imagine a technological and student-driven learning process taking place outside the classroom that is then supplemented with exercises and other reinforcing activities inside the classroom. For instance students learning neuroscience could watch videos of lectures and animations of synaptic transmission at home as well as using interactive software that determines their level of understanding. The software would be individualized and tailored to the student's particular learning style, using visual examples if they are good visual learners and reading assignments if they are better verbal learners. Teachers would access this information to identify the student's weak points and focus on those areas of difficulty to ensure comprehension. The students could then participate in lessons where they must apply what they have learned or generalize it to other domains. This technique is sometimes called "flipping" the classroom, since much of the basic learning process takes place at home.

The trend toward utilizing technology is present in massive open online courses (MOOCs). These are classes that can be taken online. They utilize both traditional activities like readings and problem sets but also provide interactive user forums where teachers and students can communicate and assist one another using email and chat rooms. Many universities are now using MOOCs at least in part because they are profitable. They have been the subject of much criticism in the field of education. Critics argue that digital literacy is a necessary prerequisite and that not all students have such skills. It also requires a lot of time and effort both for teachers and students. Participants must assume a lot of responsibility, keeping up with the readings and being mostly self-motivated.

Technology will play a huge role in education for the poor in the future and there are numerous examples of successes in this area already. The Indian physicist Sugata Mitra found that children from poor backgrounds scored as high on tests as those from top schools when they were encouraged to learn by an older student. These learning gains were obtained with just a small amount of encouragement, what he calls "minimally invasive education". He has goals to set up thousands of computers with benches around the country in impoverished areas to create self-organized learning environments (SOLES).

Another story comes from the Indian man Salman Khan, who made educational videos to teach his cousins. He uploaded these to YouTube where they went "viral" and became immensely popular. He has since founded the Khan Academy to produce additional videos on all sorts of subjects ranging from biology to art history. These are now watched by millions of people from around the world. The videos are self-paced, allowing the viewer to stop whenever they desire and to chart their mastery of a subject. A school in North Carolina has set up a learning program using the Khan videos. Students watched them at home then did homework and problem solving exercises based on the material in class.

GAMING

Video games are much maligned these days and blamed for everything from rape to murder to the decline of morality. What fewer know is that there are significant benefits of gaming. In the attentional blink phenomenon, people miss the second occurrence of a visual target item because a previous one distracts them. One perceptual study showed increased performance on the attentional blink task in participants who played first person shooter games, those in which a person has to move through an environment and respond rapidly to events like a soldier popping up from behind an obstacle. These perceptual skills could potentially save someone's life while driving.

Games are fun. They are intrinsically interesting and we want to play them. The same cannot be said for most homework and other school assignments. So it would make sense to use games to get students to learn. Katie Salen, a game designer and educator has done just that. In 2009 she created a program called Quest2Learn at a public school in New York City. The curriculum mirrors the design principles of games. Learning objectives are "missions" that require strategies like collaboration and role-playing. Hands-on problem solving exercises and game creation are also incorporated into the curriculum.

Games may also fend off mental decline. Researcher Adam Gazzaley uses a game called Project: Evo to test and challenge cognitive skills in the elderly. In Project: Evo, players guide a rocket-powered surfboard along a river while responding to the presence of colored animals like red fish or blue birds. The game is used to measure visual-motor tracking, selective attention and working memory. Gazzaley found that elderly people who played a prototype of the game showed improvements in these skills, but only when they were challenged to multitask by executing the steering and response components of the game at the same time.

Jane McGonigal, a game designer and member of The Institute for the Future has an even grander vision of what games can do. She believes gaming is a vast untapped resource that can be used to solve many of the world's problems. She cites several statistics about the amount of time people spend in games. It turns out people around the world spend about three billion hours a week playing online games. To illustrate this, gamers have spent 5.93 million years playing the interactive fantasy game called World of Warcraft. The average person in a gamer culture has spent 10,000 hours playing online games by the time they are of 21 years of age. This is the amount of time it takes to become accomplished at many real world professions. Much could be done if we could direct this time and expertise toward concrete problems facing society.

People are passionate about games says McGonigal. They involve us emotionally and bring out the best in us. We collaborate and cooperate while playing and don't get frustrated in the same way we do in real life. Instead we persist and strive harder to overcome obstacles. The ability level of games are aimed to match our actual skills so the challenges are tough but not unwinnable and we can improve as we go along to take on more advanced levels. We also achieve a great sense of satisfaction after making "epic wins", where we succeed against what appear to be unstoppable odds.

In 2007 Jane and other developers created an online game called World Without Oil in which players had to survive an oil shortage. They were given fictional but realistic information on the cost of oil and food, how transportation was affected and if there were school closures or rioting. Players then needed to go about seeing how they would live their lives under these circumstances and to blog about their experiences. The developers tracked 1,700 players for a period of three years and were able to glean a good understanding of how the world might be affected in such a scenario.

In another game the researchers told players there was only 23 years left on the planet before disaster and asked them to invent the future of energy, food, health, security and the social safety net to prevent it. They did this in a game called Superstruct. This time 8,000 people played and came up with 500 creative solutions. The solutions can be viewed by googling the word "Superstruct". In a third instance McGonigal helped create a game in which players were trained in social innovation skills like knowledge networking, sustainability, vision and resourcefulness. Done in conjunction with the World Bank Institute (WBI) and universities in sub-Saharan Africa, the players could be certified by the WBI as social innovators. The goal was to encourage people to get involved in solving problems in those countries and elsewhere.

INFORMATION PRESERVATION AND PRIVACY

We are awash in an ocean of information. All around us in the modern world are countless data sources competing for our attention. Think of how much information there is on the Internet, cable TV and your computer hard drive. What will happen to all of this? Will it be preserved for prosperity or will parts of it be lost forever in a sea of bits? The issue is a non-trivial one. As information technology evolves our way of storing and accessing information changes. Just think of music being encoded first on records, then tapes, then CDs and now as digital files. What music was lost in each transition? Perhaps more than we think.

Information is the medium of the future. It is vital that we preserve data as it is important to problems we need to solve. The tragedy is we might not even know what we are losing. Biologists may fail to discover a new cure for a disease because of a journal that went out of print. Economists may fail to understand the dynamics of the stock market because a crucial database was deleted. Artists may not be inspired to create a new style of painting because they were unable to view paintings by other artists online.

This situation has already happened in our history and the results were tragic. The Library of Alexandria in Egypt contained untold centuries of information from multiple cultures before it was burned and destroyed in several attacks. Just think what we would know about the history and discoveries of those ancient times if it had survived. Learning then and now is slowed because of this loss. Knowledge must build on knowledge. Without a proper foundation to sit on it takes longer to accrue.

A solution to this problem may be in the form of a Library of Humanity, a modern version of the Library of Alexandria dedicated to saving the history of our kind. It would be the job of a Society for the Preservation of Information to determine which knowledge is crucial, how it could be rescued and then preserved. This may require the creation of a generic information code, a universal way of encoding, storing and decoding data in a way that could be figured out by a sufficiently intelligent species or by our descendants in the far future.

With the information explosion privacy has been at the forefront of the news lately. In the U.S. we know the government has been listening in on our cell phone conversations and that corporations are monitoring our E-mail and Internet traffic. Understandably this has upset many people. We need laws that balance information privacy with national security. A commercial information service provider such as Google should not be able to provide data about its clients to government or other parties without consent or a warrant. We need to ensure that these providers delete data like our search terms and buying preferences or at the very lease hold them for a specified period before deletion.

Privacy is getting harder to come by these days. When we step out of our homes we can be monitored and recorded in all sorts of ways ranging from public security cameras to airborne drones. Facial recognition algorithms can now recognize us from great distances and in noisy backgrounds like a crowded pedestrian mall. Privacy in our own homes and on our own property ought to be sacrosanct, but when we are in a public setting we must acknowledge that we cannot have perfect privacy.

Since technology is driving much of this change it can also offer a solution. Radar detectors and white noise generators are already on the market and available to consumers. In the future we may be able to install electromagnetic band jammers like those used by the military to cloak a specific location such as a house. Since drones are getting smaller and more popular there will inevitably be a new set of laws that prohibit them from operating in dangerous areas and that will require operators to register and obtain a license.

Another recent concern is hacking of commercial institutions. Several major banks and retailors have recently had their systems compromised by criminals. These thieves were able to access vital consumer information like credit card numbers. Information in these accounts needs better safekeeping. The future is likely to see a tremendous increase in the science of encryption. This will likely be a seesaw battle between hackers implementing new techniques and developers protecting against them.

One of the nice things about the World Wide Web and the Internet is that they are distributed systems. In these systems data is stored and processed on different servers across the globe. An attack on one part of the network does not completely cripple its global functionality. But there are weaknesses. Some networks are hub-based, meaning they have one or more central locations receiving and projecting large amounts of data traffic. Taking out a hub can result in partial network outages on its subsidiary nodes. New developments in the field of network science can better prepare us to defend against these kinds of attacks.

Given the importance of information to our daily existence it may be worth taking a novel approach to how it is valued. One approach is to create an information commodity exchange in which information is publically traded. In this system participants would buy and sell shares representing different ideas. Formulas for a new drug, alternate architectural plans for a proposed building or designs for a new computer could all be assessed this way. The value of an idea would be directly proportionate to the value of its stock. The system is prone to the same problems as the regular stock market, with over and under valuation possible but it would at the very least provide a metric by which to judge an idea's quality.

CHAPTER 2. NEW SCIENCE

QUANTUM MECHANICS

Quantum mechanics is a field of physics used to describe small-scale phenomena, those at the atomic and subatomic level. According to quantum theory, certain attributes of particles such as their position or momentum cannot be known with certainty. Instead they are best understood using probabilities that describe the chance that, for example, a particle is located in a certain region of space. This probabilistic approach poses problems for how well we can understand small-scale phenomena. According to the Heisenberg uncertainty principle the more precisely you know the position of an atomic particle the less precisely you can simultaneously know its momentum. This principle in effect states that we are limited in how well we can understand quantum behavior.

What is fascinating about the quantum approach is that it implies some events don't actually exist or fail to exist. They are in a state of indeterminancy or probability until being observed, at which point they are made certain. This combination of states is called superposition. The classic thought experiment about this is Schrodinger's cat in which a cat is locked in a box with a radioactive atom, a Geiger counter and a vial of poison gas. If the radioactive atom decays it is picked up by the counter which then breaks the vial, killing the cat. If no decay occurs the cat lives. In this scenario the cat is neither alive nor dead after one hour. Once the box is opened, the observer forces the cat to assume one of these two outcomes.

Quantum mechanics posits that light can be both a particle and a wave. The light itself is made of particles called photons but the wave is the probability distribution of where those particles are located in space. One way to think of this is by throwing darts at a dartboard. After 100 throws the locations of the darts stuck in the board are the particles but the pattern of darts across the pad corresponds to the wave. So light can exist as a collection of particles but its behavior is best described as a wave.

Another bizarre aspect of quantum mechanics is quantum tunneling. When a probability wave hits an energy barrier such as a wall most of the wave is reflected back. But there is a chance that a small amount of it will "leak" into the barrier. Should the barrier be small enough the wave can then continue on the other side. In effect, it has tunneled or travelled through the wall! This can only happen at small scales. If we throw a basketball against a brick wall it will not somehow magically appear on the other side.

So our understanding of how everyday objects operate goes out the window when dealing with quantum level phenomena. It states we cannot measure things perfectly, that observers actualize events, that things can exist like particles but act like waves and that sometimes particles can mysteriously tunnel through walls. However all of these observations are supported empirically and taken as true by the modern physics community.

Quantum mechanics poses a problem because the rules that govern it are not the same as those used to describe things at larger scales. Newtonian physics is best at depicting how things work at the scale we all live in: the world of people, buildings and trees, etc. It was the first to describe the physical motion of objects. Einstein's physics is best for describing what happens at very large scales and can account well for things like gravity, space and time. So we end up with three different physics: quantum for the microscopic, Newtonian for middle scales and Einstein's physics for the macroscopic. We have yet to develop a theory of physics that encompasses and explains all three with a unique set of equations. This is one of the holy grails of modern physics.

It is worth digressing a bit here to talk about scale. Most of the time when we think about space, it is in terms of distance. We think how far we need to travel to get from point A to point B or of how long it will take to get there going at a certain speed. Less often do we think of space in terms of the really small or really big.

A great online animation to get a sense of scale in the universe can currently be viewed at: htwins.net/scale2/

The immense differences in scale beg interesting questions. Is there an upper limit on how big things can get? We don't know how big the universe really is, since we are limited to measuring only what is observable. Likewise we don't know how small things can get. Particle accelerators like the large hadron collider that smash small particles together at high speeds are one way of doing this. Why do we see life only at a particular range in scale, from the virus (approximately 10^{-6} meters) to the whale (about $10^{1.5}$ meters)? Is there something special about this range? These questions are only complicated by the notion of the multiverse: that there may be multiple universes that branch off from each other.

Just because we don't have an understanding of the really small however doesn't mean that we won't. Also, we have been able to apply our understanding of the quantum realm to create things like quantum computers. Rather than using a Boolean system of zeros and ones as an ordinary computer does these devices instead encode information as qubits that can exist in superposition as a zero, a one or as any state in-between. This gives them the potential to be millions of times more powerful than the computers we are using today. There are only very simple versions of these now, but more complex ones are under development.

It is difficult to accept quantum mechanics and its applications because the concepts don't match our everyday experience. This is true for many findings in science. In the essay on space travel we see how difficult it is to comprehend the vastness of the universe because the distances are just so great. The same thing is true for time. So much time has elapsed since the start of the universe that we cannot think of it meaningfully given our short lifespans. This is one of the limitations we must confront in the future. If we can't relate personally to these concepts it may be difficult or impossible to reason about them effectively. One goal for humanity then would be expand our own intelligence in ways that would enable us to think in terms of unfamiliar concepts and large scales.

Mathematics is a step in this direction. Perhaps through genetic engineering or cybernetic computer implants we may be able to achieve this. Solving the mysteries of physics and cosmology may depend upon it.

COMPLEXITY

The brain and other phenomena like the stock market and the weather are examples of complex systems. A complex system is one whose properties are not fully explained by an understanding of its component parts. There are several features of such systems. They are very sensitive to changes in initial conditions. Alterations in starting conditions propagate through the system, producing unanticipated outcomes. An oft-cited example of this is the butterfly effect. In this scenario, a butterfly flaps its wings in Brazil producing a tornado in China or some other far-removed location. In a complex system there are multiple interactions between the many different components. The parts affect one another in an intricate causal dance that is hard to track.

But what exactly do we mean by complexity? What makes one thing simple and another thing complex? There are several different approaches we can take. One way to look at this is in terms of order. Simple systems are ordered. As such they can be broken down into their component parts and part interactions to understand how they function. Examples that fall into this category are machines. If we reduce the amount of order in a system we get into the realm of ordered complexity or chaos. Systems here are characterized by a combination of randomness and order. Sometimes they act in ordered ways and sometimes they act in ways that are unpredictable. Biological systems like our own bodies fall into this category.

Then there are those systems that are completely unorganized. These have no order and can only be described statistically. We can think of these as aggregates of items or particles such as air molecules in a volume of space. A reductionist approach to these systems breaks down and we can only express their behavior probabilistically. Entropy is a measure of a system's disorder and is one way to quantify these systems. Measures of entropy would be low for ordered systems, at intermediate values for ordered complexity and high for aggregates.

Another way to think about complex systems is in terms of algorithmic information content. In this view a thing's complexity can be described by using the fewest terms in a formal description. For instance, the string 00001111 is redundant. It has many of the same items repeating. It could be represented as 4(0), 4(1). The string 01001110 on the other hand is more complex. It requires more terms to describe: 0, 1, 2(0), 3(1), 0 and is less compressible. These sorts of descriptions come from a field known as information theory.

There are also systems that are computationally complex. In the traveling salesman problem a person needs to travel the shortest route connecting a given number of cities. For a small number of cities the possible ways of doing this are relatively straightforward. But as we increase the number of cities that must be visited the number of routes increases dramatically. For twenty-five cities, the number of possible routes is so large that even a computer searching a million possibilities a second would take 9.8 billion years to look through them all! Problems of this sort are intractable.

Complex systems are difficult to predict, understand or control. If the human brain and body are complex systems then we may never be able to figure them out. One response to this is that just because a system is complex, doesn't mean it can't be simplified or understood. One way to think of a complex system is as a tangled plate of spaghetti. Each strand represents a causal influence. At each point where one strand contacts another it exerts an effect. When we step back and look at the plate of spaghetti, it appears as a tangled unexplainable mess. But if we start to measure where the strands touch or where they start and end, the situation becomes more comprehensible.

In the future we will be confronted with problems of great complexity and we need to be ready to solve them. Fortunately there are organizations already in existence that are applying complex systems theory to a wide range of concerns. The New

England Complex Systems Institute is developing novel approaches to the problems of economics, ethnic violence, health care and education among other areas. They combine methodologies from physics, computer science and mathematics with computer simulations and high dimensional data analyses to describe real world patterns of behavior. The Santa Fe Institute in the U.S. and the Centre for Complexity Science are doing similar things.

SELF-ORGANIZATION AND EMERGENCE.

Self-organization is a process by which the internal organization of a system increases in complexity without being guided or managed by an outside source. Self-organizing systems typically display emergent properties. Emergent properties are those that cannot be explained by the sum of their parts. They introduce a new level of complexity or behavior that seems to go beyond what the system is capable of.

According to Camazine there are several features of organized systems. First, they are dynamic. They require continual action to produce and maintain order. However they are also stable. Biological systems such as ecosystems are stable. They tend toward conditions of constancy unless acted upon by an outside force. Also they are mostly dependent on local interactions, those that are happening within the system itself. One of the unique features of these systems is the ability to generate complex behavior from simple parts and part interactions.

In self-organization, the pattern at the global level emerges solely from interactions among the lower-level components. The rules specifying the interactions are executed using only local information without reference to the entire pattern. There are numerous examples of self-organization in the physical, biological, and human realms. These include crystal growth, chemical autocatalytic sets, bird flocking, cellular automata and various phenomena in economics.

Animal bodies including our own are self-organized. They are hierarchical, with levels nested one inside the other. Atoms make up molecules, molecules make up cells, cells make up organs and organs make up bodies. At each level we see new and more complex behaviors emerge. Molecules by themselves have a given repertoire of behaviors based on their properties. But if we put a group of them together inside a membrane they begin to interact and produce all sorts of emergent behaviors such as protein synthesis and the Krebs cycle.

Nature uses self-organization to build things. There is no central coordinator who tells the system what to do. Neither is there a comprehensive blueprint or set of instructions that can be followed to put the elements of a system together. These are examples of top-down organization in which information is known ahead of time. Human societies build things this way but nature does not. Nature is bottom-up.

Let's take as another example the development of the brain in a human fetus. Neurons "know" which cortical layers to migrate to by climbing up radial glial cells much the same way a person might climb up a pole. The glial cells that are already in those positions serve as a local "construction crew", assisting in the building of the brain. So there is a sequential dependence in self-organized development. The results of a previous process contain the information necessary to carry out the next.

Most complex systems as we have seen are characterized by having organization at different levels. The brain can be studied at the level of a neuron, neural circuits or brain regions. A society can be studied at the level of the individual, family, school or state. Each of these levels may be thought of as emerging from the elements below it but as also possessing new properties and behaviors not shared by the levels beneath it. A scientific study of complexity requires that we be able to define formally what we mean by an emergent level. If this were possible, we could analyze how levels form, examine what levels share in common with each other, and compare levels across disciplines.

There have been attempts to do this. Goldstein suggests we carve systems at their "joints" and define emergent levels as places where we see differences in organization, characteristics and behavior. Different science disciplines are defined this way.

Chemists study chemistry of atoms and elements. Biologists study cells, organs and bodies. Psychologists study the brain and behavior. Sociologists, economists and political scientists study social phenomena. The breaks between these disciplines are where the levels appear.

DYNAMICAL SCIENCE

Science is our most powerful tool for understanding the natural world. It has only been around a few centuries but in that time we have learned more than in all the time that came before. In fact, we probably generate more new knowledge in one day now than may have been generated in an entire century earlier in our history. There is still a lot left to know but in a relatively short period we have determined much concerning the physical rules that govern our universe, the chemistry and biology that governs living creatures and the large-scale mapping of the universe. We have unlocked our genome and are on the way to developing the connectome, a complete, cellular level wiring diagram of the human brain.

Science despite its successes has some limitations. Researchers tend to focus on linear phenomena. These are relationships between variables that are characterized by simple increases or decreases. Nature is mostly nonlinear, meaning outputs are not simply proportional to inputs. For example a plot of a person's happiness over time will go up and done repeatedly in a complex way.

There is also an assumption in many of the sciences of independence, which is assuming that the parts of a system are compartmentalized and not influenced by other parts. Some cognitive scientists treat the brain this way assuming there are "modules" that process inputs autonomously. This is a simplification and helps us to understand the complex process of mind but is not how the brain actually works. At best we can say the brain is quasi-modular and that many natural processes are characterized by interdependence.

An advantage of making an assumption of independence is that the system can be understood analytically. In this approach the phenomenon in question can be understood by putting the separate descriptions together. This approach however often fails. Many systems in nature resist reductionist explanations because they function in a holistic manner. A system is holistic if cannot be broken down piecemeal. Each piece relies on the parts around it to function properly. In the words of the Gestalt psychologists, "the whole is greater than the sum of its parts". Behavior in this account can emerge from the interaction of the parts spontaneously.

The dynamical systems approach, sometimes also known as chaos theory or complexity studies, takes these considerations into account. In this perspective systems are nonlinear, interdependent and holistic. In addition, phenomena are assumed to be multivariate where there are multiple forces at work that shape what we measure. They are also multivariate with regard to outcomes and so the influence of many variables is studied in terms of their influence on many other dependent variables. Oftentimes these effects are plotted in what is called a state space that shows the way a system changes over time.

One of the most vexing questions concerning the dynamical systems approach is that the behavior of a system can be deterministic yet non-predictable. By determined it is meant that there is an equation that specifies the system's behavior that allows for an accurate specification of input and output variables. Yet at the same time the system is non-predictable in the long run. How can this be? If we understand something completely we should also be able to predict it completely. The non-predictability of phenomena like weather or the human brain means we really don't understand them because if we did our equations would be able to predict what happens at any point in the future.

There are several reasons why science has yet to adopt the dynamical paradigm. It requires not only a different conception of the world but new statistics and data analysis techniques that require researchers to retrain. It can be difficult to keep track of multiple variables, as the relationships between them are complex.

Methods and paradigms in the sciences can be slow to shift and even though the dynamical approach has been around for some time, only a few science practitioners actually adopt it.

A number of researchers claim that science lacks unification. Among these proponents is E. O. Wilson. In his book Consilience: The Unity of Knowledge, he claims that the different fields of science are fragmented. They posit different causal and explanatory factors. One theory that could unite them is the evolutionary approach, according to which selection pressures act on variability in a population to determine which genes get passed on.

Another technique that may be utilized more in the sciences is computer modeling. Meteorologists use modeling to forecast the track of hurricanes. Chemists use them to determine how amino acids fold up to form three-dimensional proteins. Cognitive psychologists sometimes use them to determine how the brain might perform some process. Particular to this last field are the use of artificial neural network (ANN) models, which are software programs designed to mimic the brain. ANNs have been used successfully to model language development and acquisition of arithmetic skills.

In the future as greater computational power becomes available we will likely see more and more complex ANN models to account for more complex cognitive and neural tasks. The Blue Brain Project started by a Swiss research group is even more ambitious and plans to model large portions of rat and then human cortex containing millions of cells down to the molecular level. Other computational models have simulated the way the hippocampus, the brain area responsible for memory encoding, operates.

Computer modeling is not without its critics. One danger is that the models take into account too few variables and thus fail to adequately represent real world complexity. Increased computational power through the use of supercomputers can partially solve this problem. A second issue is the extent to which models can make accurate predictions. To maintain predictive accuracy the models need to be tested repeatedly against empirical data. They should always be used in the context of experiments. If a model fails to predict data generated from an experiment, it must be modified until it does. In this way modeling and experimentation can work iteratively together to produce a better understanding.

NETWORK SCIENCE

Recent years have seen the development of a new area of study called network science. Scientists, mathematicians and others in this area have begun to explore the way complex networks operate. A network in this view is considered as any collection of interconnected and interacting parts. Like cognitive science, which is made up of investigators from multiple disciplines, network science is also interdisciplinary. Investigators come from different areas like mathematics, physics, finance, management and sociology. This is because networks are everywhere, as electromagnetic fields, stock markets, companies and societies.

Contemporary network scientists additionally consider networks as dynamical systems that are *doing* things. For example power grids distribute power, companies manufacture products, and brains think. All of these are networks in action. The structure of networks can also change over time. New links can form and new organizations can emerge. The new science of networks investigates not just network structure but also its function. Researchers in this field are interested in the anatomy of networks, how they operate and how these operations evolve and change.

Network scientists consider networks as abstract structures and to some extent they can disregard their specific context. Whether the network is a brain, a traffic system, or the Internet in some sense doesn't matter. That is because there are commonalities these networks share. In other words, all networks share some universal mechanisms of action. For instance, most known networks exhibit a critical point where activity propagates suddenly through the entire system. It seems amazing that cars on the road and employees in a company should act the same, but because they are both elements in interconnected relationships, they do.

It is also important to note that although networks do share some characteristics, they don't act exactly the same. In some cases, the characteristics of the nodes that make up the network produce different behavior. A neuron in the brain acts differently from a consumer in a society. So we need to acknowledge that there are both universal and particular features to networks.

One important question that arises when dealing with networks is how information in them is coordinated. This topic is known as centrality. In human societies there is a leader who can perform this function. In a computer it could be considered the central processing unit (CPU), but in brains there is no "leader". Human brain activity is characterized by simultaneous activity in different regions. It is likely that neural synchronization is what unifies these disparate processes. In neural synchronization, neurons in the frontal lobes begin firing at the same rate as neurons in other parts of the brain like the occipital lobes, perhaps unifying the information and allowing it access to conscious awareness.

An additional feature of brains and other networks is hierarchical organization where there are different levels of processing. In the visual system neurons at one level respond to spots of light. Those then feed to the next level in which line representations are formed. Those in turn feed to another level where angles are represented. As one travels up the network, the features become more and more complex until at some point there are neurons that seem to stand for entire objects. Hierarchical organization is also seen in the military and in corporations. In these groups people of differing rank are responsible for issuing orders to those below them. Communication in hierarchal networks like these can be two-way: messages can travel up and down the levels.

Small-world networks are an exciting and recent discovery. These are networks where messages can be transmitted rapidly between any two parts of a network. These architectures are characterized by short-range local connections as well as long-range global connections. The latter allow information to propagate between widely disparate regions, bypassing "local" traffic. It is this organization that allows for neural synchrony. It is also the mechanism by which epilepsy can spread from one hemisphere to another and by which disease can spread across continents.

Network science is a powerful framework that we can use to better study cognitive and other natural processes. Researchers in this approach have uncovered anatomical and functional features that underlie many systems whether those are brains, transportation networks or ecologies. The application of network principles should yield some fascinating insights in the coming years. A computer built using a massively parallel architecture coupled with a small-world network organization might result in a conscious machine.

CHAPTER 3. NEW TECHNOLOGY

TECHNOLOGY AND THE FUTURE

Perhaps the most important issue facing our future is technology and the role it will play in human society. We have seen vast progress in technological change in recent years. Moore's law since its inception still holds, with a rough doubling of computing power and memory every 18 months. Up until the last few decades this exponential increase in technological progress was unheralded in all of human history.

We seem to forget how tough things used to be not so long ago. In 1918 there was a flu pandemic that affected about 30% of the worldwide population and killed an estimated 50 million people. Improved medical care has been able thus far to contain such outbreaks. Imagine what your life would be like without most of today's amenities. How would you get food without a supermarket? Clean your clothes without a washing machine? See at night without electric lights? Keep cool without air conditioning? It is difficult to imagine what daily life must have been like only a century ago. Much of people's time back then was spent in menial labor just to take care of life's essential needs.

The fact is that technology has freed up our time and allowed us to be more productive at work and at home. Not only this but we now have many more forms of entertainment. Not so long ago, reading a book or listening to live music were some of our few entertainment options. Now we have radio, television, video games and the Internet all at the touch of a button.

One of the concerns brought up by technological development is its destructive potential. We are now faced with the danger of weapons of mass destruction like atomic bombs and biological agents that if in the wrong hands could spell the doom of millions or maybe billions of people. Technology is inherently neutral though and can be put towards good or evil ends. This is a human problem and one that must be solved at the human level. We need to curb our aggressive tendencies and fundamentalist beliefs.

Another concern about technology is how we as biological beings interact with it. Computers over the past several decades have not only gotten faster but smaller. They went from occupying rooms to desks and from there to laps and palms. The next wave of technologies will be "wearables" and will be embedded in our glasses, clothing and other accessories. Following this, computing devices will be embedded in our bodies and the difference between biology and technology at that point will begin to blur. Possibilities include retinal implants that project visual information and neural prosthetics that will allow us to control devices remotely just by thinking about them.

I think this gradual integration of tech into our bodies actually answers one of the fears we have about technology. The fear is that computers and robots will become more intelligent than we are, deem us unworthy and promptly begin to wipe us out. This view sees computers and robots as entities separate from us. In the future biology and technology will be so intricately intertwined that any attempt on "their" part to kill "us" will kill all of us. We will be so entangled and interdependent that anything detrimental to humans would also be detrimental to machines. A phrase that expresses this is: "They will be us and we will be them".

So when it comes to technology we must be careful not to "throw out the baby with the bathwater". Its contribution towards increased quality of human life has far outweighed its actual and potential dangers. In the future we will need to focus on how technology and its human operators can go wrong and implement preventative measures. Some solutions include a reduction in aggression and fundamentalist beliefs on the human side and fail safe systems and other security measures on the machine side. See the essay on friendly AI to learn more about how we may contain dangerous computer programs.

TECHNOLOGY AND WORLD PROBLEMS

Peter H. Diamandis and Steven Kotler have recently completed a book titled *Abundance: The Future is Better Than You Think*. In it they outline the world's basic needs and how those can be met using current technology. They posit a world pyramid with basic level needs at the bottom and increasingly more sophisticated needs going up to the tip. This is actually a version of Maslow's hierarchy of needs applied to society. Maslow's pyramid is for individuals and is discussed in the higher order needs chapter later in the book. At the bottom level of the world's needs pyramid are needs for clean drinking water, nutritious food and proper housing. At the next level up are energy, communications and education needs. At the highest level are health and freedom needs.

The authors provide numerous examples of how these problems can be taken care of by innovative technology. For water needs, the entrepreneur Dean Kamen has created a water filter called "Slingshot" that produces clean water with very little energy use. For food needs they cite genetic engineering that has increased food yields to accommodate an increasing world population. Hydroponics is also a very promising development, as this type of agriculture does not require fertilizer or pesticide. Cell phones have had a transformative impact in countries like Zambia, where the poor can order items without bank accounts. Advances in health have been made using a "lab on a chip" that has been used to test urine and blood samples in the field and so perform diagnoses without the use of an extensive medical infrastructure.

Interestingly, many of the innovations Diamondis and Kotler cite are not the product of government or industry but of enterprising individuals. They describe three groups of people who have been making such changes. The first are the "Do It Yourselfers" (DIY). An instance of a DIY invention is Chris Anderson's quad copter drone that is a major improvement upon existing models and is in development for delivery of goods (Who wouldn't want beer delivered to you while ice-fishing?). Then there are the techno philanthropists. These include people like Bill Gates with his Bill and Melinda Gates Foundation and Nicholas

Negroponte's One Laptop Per Child (OLPC) program. The OLPC program has produced a $75 tablet that has been distributed to third world countries to stimulate learning and education. The final group they refer to as the "rising one billion". This refers to the global population of people living in poverty. There are surprising stories from this group like Arvind Mills, a successful Indian businessman who started off having local tailors make jeans. Microloans to the poor have also shown some successes in stimulating local economies in disadvantaged areas.

Diamandis and Kotler argue that incentives are necessary to stimulate invention. This is the impetus behind The X Prize, a monetary award for an invention that has the potential to change the world. The X Prize has to be winnable by a small team, achieved within a reasonable time frame and be able to induce others to invest and compete in the same area. Among the X Prizes in effect now there is a $15 million dollar prize to empower children to take control of their own learning in poverty stricken areas around the globe, a $2 million dollar prize to help heal the Earth's oceans and a $10 million prize to create a handheld medical diagnosis device. Past prizes have already been awarded several times and often won by underdogs.

We are in a shrinking world where airline travel and mass communication allow us to solve problems cooperatively without having to be physically next to each other. Examples of such cooperation abound. A Canadian and an Indian company have worked together to create a $35 Android tablet. People making films together on YouTube generate more content in 25 minutes than Hollywood does in a year! We also need to be open to new techniques that allow us to be creative. Michael Schrage, an M.I.T. researcher developed what he calls the 5*5*5 method of business innovation in which 5 teams of 5 people each have 5 days to come up with 5 business plans that do not cost more than $5,000. There are numerous other techniques possible but few of them have been tried, perhaps because most people are afraid to think differently and challenge the status quo.

Diamandis and Kotler note that we should be upbeat about the future. Things are getting better, not worse. As mentioned here in another essay, there are now fewer people dying in wars, less crime and increased liberalization with regard to human rights. Despite this, most people have a gloomy outlook about the future. Why is this? One reason is psychological. We tend to be more optimistic about what we ourselves can accomplish. This is because we can imagine ourselves doing better and have some degree of control over our own actions. With others it is just the opposite. It is harder to imagine what a beneficial future might be and most of us feel we have little or no control over how that could transpire.

The news media are also responsible for this gloominess. Studies have shown that the news is up to 90% negative. This is done to get our attention and keep us watching. The situation has gotten so noticeable that some news stations actually have a section devoted entirely to good news. Mass communication also plays a role here. Any bad news no matter what its scale and regardless of its location is picked up and reported. One hundred years ago you would never have heard about a double homicide on the other side of the country. Today it is commonplace. Also, many reported threats never materialize. Anyone remember the killer bees? Y2K? Acid rain?

GENETIC AND MOLECULAR MEDICINE

The Futurist Michio Kaku mentions three stages in the evolution of medicine. The first lasted for tens of thousands of years. It was characterized by superstition and witchcraft. Many treatments during this time had no scientific basis and only exacerbated conditions. However, some treatments in ancient Chinese medicine like herbal remedies and acupuncture have been found to be beneficial, probably through a process of trial and error. The second stage started in the nineteenth century when we understood that germs were the cause of many diseases. This lead to greater sanitation and a corresponding increase in expected lifespan. The last stage we know of, the one we are in now, is molecular medicine and involves an understanding of medicine in terms of atoms, molecules and genes.

The Human Genome Project is an international scientific project aimed at determining the entire genetic sequence of human DNA. The goal is to map out the nucleotide base pairs for all of our genes and to understand their role in development and physiological function. A gene is that portion of our DNA that codes for a protein. Nucleotides are the pairs of molecules that constitute the code. They go in pairs, adenine with thymine (A-T or T-A) and guanine with cytosine (C-G or G-C). Proteins are the little molecular "workers" of our bodies. They play an important role in orchestrating all sorts of chemical reactions. The first attempt at human genomic sequencing was successful but very expensive, costing three billion dollars. The goal now is to perform genomic mapping for individuals at a substantially lowered cost, under $1,000. Another related field of study is bioinformatics. Researchers in this field attempt to determine the genomes of many different organisms, which will allow us to understand the evolutionary relationships between them.

Genes are implicated in disease. For example there are certain genes which if present in our DNA increase our likelihood of getting a disease like Alzheimer's. Alzheimer's disease is a degeneration of the brain resulting in decreased cognitive function and ultimately death. Assuming low cost genomic mapping in the future, we will be able to determine our risk factors for different diseases. This is a double-edged sword. It enables us to practice healthy behaviors that can reduce our health risks, but it is also very worrisome and can produce a negative toll on health in terms of anxiety. People will probably have different reactions to knowledge about their genome. Some will refuse to know anything and prefer being in the dark. Others may wish to learn about certain parts of the report, only opting for news of those genes for which pre-emptive behaviors can significantly reduce risk.

Gene therapy is the transplantation of normal genes into cells in place of missing or defective ones in order to correct genetic disorders. This is accomplished through a "vector" or virus that is used to inject the gene into target cells. There are many diseases that may benefit from this treatment. They include Tay-Sachs disease, cystic fibrosis and sickle cell anemia. This technique is still in its infancy though and there is much work that remains to be done before effective treatments appear. A number of clinical trials are taking place in the U.S. and Europe to evaluate the effectiveness of gene therapy.

Stem cells are "proto" or generic cells that have yet to specialize. During embryonic development in the womb cells start this way but eventually migrate into different locations to become epithelial tissue for the skin and nervous system tissue for the brain and spine, to list two examples. They have great potential in treating a whole host of problems including Alzheimer's disease, Parkinson's disease and cancer. Animal studies have utilized stem cells to successfully regrow damaged spinal tissue. There are ethical dilemmas to consider here, as many object to the use of embryonic tissue. There is also the danger that stem cells will multiply and proliferate out of control.

Suppose you are in an accident and need a new liver. There are no compatible ones available and you could die unless you get one soon. Not to fear however, as the year is 2050. In your childhood several tissue samples were taken from your body. One of those samples is now used to grow an entirely new liver organ that is then implanted. Since it was grown from your own tissue there are no compatibility problems. This procedure is called tissue engineering. Cells from the earlier biopsy are placed into a mold and bathed with growth factors. This stimulates the cells to multiply and fill up the space inside the mold. The surrounding scaffolding is then removed and when implanted into the body, resumes its normal physiological functioning. The near future may see the use of this procedure to repair cartilage and bone and to promote proper dermal wound healing.

We describe nanotechnology in greater detail in the next essay. It also has tremendous potential to cure many health ailments. Imagine thousands of tiny nanobots injected into the body. These are small artificially synthesized molecular machines. They could float through the blood, programmed to perform many functions. For instance they could attack and destroy cancer cells or viruses such as AIDs and Ebola. Alternatively they could be programmed to release insulin as a treatment for diabetes or even serve as artificial red blood cells, increasing the amount of oxygen we can absorb.

NANOTECHNOLOGY

Nanotechnology is the construction of very small objects and machines. They are assembled on the atomic or molecular level. We have reached the stage now where we can actually move individual atoms to build things! These scales are so small that quantum effects need to be taken into account. A nanometer (nm) is one billionth of a meter. To give you a sense of scale, the bacterium Mycoplasma is around 200 nm in length, the DNA molecule diameter is about 2 nm and small atoms are about one fourth of a nanometer. This field has much potential and could lead to great advancements in medicine and electronics. As such it has been funded to the tune of billions of dollars in the U. S., Europe and Japan.

It is possible to construct molecular objects using principles similar to those found in nature, utilizing the natural tendency of certain atoms to want to bond with each other. This process is known as molecular self-assembly. More difficult in some ways is creating miniature versions of devices that resemble larger mechanical versions with gears, wheels, switches and other interlocking moving parts. Examples of devices of this sort that have already been constructed are a guitar whose strings can actually be plucked and a car that can roll on wheels. Both of these objects are so small they cannot be seen with the naked eye.

Rolling up a sheet of carbon atoms in a particular way creates carbon nanotubes. These are hundreds of times stronger than steel but six times lighter. If used in the construction of airplanes and automobiles they could introduce considerable cost savings on fuel. They could also be used as transistors in microprocessors. Other products that have been manufactured using nanomaterials are clothes, packaging, paints, sunscreen and cosmetics.

One exotic far future possibility is molecular manufacturing in which nanoscopic machines called assemblers put together nanomaterials to construct nearly any type of object from food to weapons. Imagine these as 3-D printers but with accuracy at the atomic level. However it would take trillions of assemblers to make any reasonably sized object in a short amount of time. Eric Drexler hypothesizes that assemblers could first be used to manufacture multiple copies of themselves and only then sent to work creating other things.

There is also great potential for nanotechnology in the field of medicine. One proposal calls for the use of nanorobots programmed to attack cancer cells. Another envisions nanorobots as performing delicate surgery at a level of detail far smaller than any existing scalpel. Some people may (literally) have difficulty swallowing such things, especially early on when the procedures are not in widespread use. People now are comfortable putting technology in their pocket or wearing it on their wrist, but implanting devices in our bodies as with neural prosthetics or biologically active nanorobots may simply be too repulsive for some people. This fear may go away if these technologies have been in use for a while and have been proven to be safe.

There are some concerns about the safety of nanotechnology. If some of these materials are released into the environment they could cause potential harm to the ecosystem. Silver-based particles with antibacterial properties have been touted as one example. Another fear is that nanorobots could replicate endlessly and convert all matter on the surface of the Earth into a"gray goo". More realistically there is some evidence that inhalation of nanoparticles is damaging and could potentially cause cancer and other disorders. Technophiles and futurists argue that nanotechnology could potentially overcome these problems. They say we could program airborne nanorobots to rebuild the ozone layer, remove pollution and clean up oil spills.

Nanotechnology could additionally be used to alter our physical form, changing our faces or bodies at will. Imagine wanting a longer nose or fuller lips or wanting to look more attractive or like a Hollywood celebrity. This could pose a real problem for the police. How can you track someone down when they can change their appearance so radically and in a potentially short time frame?

Nanotechnology could also be used to make us smarter and stronger or perhaps give us the ability to heal quickly or see in the dark. Making those changes would convert us to the "post human" in which case we have a problem identifying who is human and how we assign rights to such beings. This could lead to two classes of humans, those who can and do enhance using the nanotechnology and those who can't. Again, this "haves" vs. the "have-nots" argument is an oversimplification and doesn't actually mirror what happens when new technology is introduced. Typically there is a lag time where the disadvantaged are unable to use new technology because it is too expensive. Then prices come down and these groups are able to afford it.

There are other economic implications. What would happen to the economy if all manufacturing could be done molecularly? This might cause widespread unemployment, but again as argued elsewhere, this is a short-term effect and is almost always followed by new job opportunities and wealth creation. Should replicators come to pass we may have no need for currency exchange since anything we want could simply be produced at the push of a button. This outcome seems unlikely as there will be costs involved in using replicators, both for the material used in their construction and production as well as the cost of manufacturing and maintaining them.

MEGAENGINEERING

Engineering is the application of science. Whereas science uncovers knowledge and determines how things work, engineering puts that knowledge to use. Technology can be considered a subcategory of engineering and is discussed in several other sections of this book. The future will probably bring with it larger and more ambitious engineering projects. Designing and building on this scale is known as megaengineering. Examples that have been proposed include floating cities, a trans-Siberian railroad across the Bering Strait and a space elevator that could link the Earth's surface with a space station above the planet.

Megaengineering projects are designed to solve some social problem on a global scale, whether that is a response to climate change, overpopulation or some other crisis. Because of their scope they will take some time to build and will be quite costly. International cooperation, corporate sponsoring and cost sharing will help. Whether such projects go into effect depends also on demand and the extent to which people are ready to support it. If costs can be managed, there is an urgent need and they can be done with existing tools and materials, they stand a better chance of being built.

Successful examples of megaengineering projects in place now include the extensive dam and water containment system used to control flooding in the Netherlands and the Three Gorges dam on the Yangtze River in China used to control flooding and to generate electricity. Dubai seems to be leading the world in projects of this sort and has constructed the world's largest artificial island and the world's tallest skyscraper, the Burj Khalifa, that stands 2,722 feet tall.

Let us examine one such project in greater detail, the idea of a personal pod transportation network that could be implemented in a place like Washington D.C. in the not-too-distant future. Pods are small, enclosed compartments that run on an elevated rail system. They can accommodate several people and operate off of electricity. A pod can be summoned from nearly any location in a city and programmed to take its passengers to other locations. The advantage of pods is that they reduce vehicle emissions by decreasing the number of cars and buses on the streets. They are meant to supplement underground subway systems that are already overburdened. Pods can come with the capability to scan for explosives or other dangerous materials reducing the risk of terrorist attack. Another perk is that they should substantially reduce driving accidents due to intoxication or fatigue.

Most people living in a city would probably welcome the addition of a pod-based transportation system. However, there are some who may regret relinquishing control of their cars. Not all locations will be accessible, as they will service primarily the inner city and immediate suburbs but not areas farther removed. Since they will be monitored with security cameras and shared with other riders, there is not much in the way of privacy. Pods are designed primarily for passenger transport and expansive storage capacity won't be available at least in current designs.

If we want to accelerate farther into humanity's future to anticipate what might get built we get to the topic of astroengineering. This field refers to the construction of machines and other structures in outer space. The International Space Station is a current example of astroengineering. There are a number of other much more advanced proposals that would only be feasible centuries from now. One is a Dyson sphere that would surround the sun and utilize all its radiant energy. Others include giant rings and disks in space that could be made habitable.

What might the psychological effects of megaengineering projects be? Some people in the presence of such structures may experience alienation, powerlessness and a diminished sense of self. These individuals may yearn for more power and control over their lives. Although the future may make things easier for us, it removes us from the "drivers seat" (literally in the case of the personal pod system or self-driving cars). Some people may look at our mega constructions and feel a sense of pride in what we can accomplish. Others, as mentioned in the section on spirituality, may feel a sense of awe and wonder. It is important for us as a species to continue pushing our abilities to their limits and beyond. If we yield control to automation in one area of our lives we need to ensure that we take on challenges in another.

CHAPTER 4. INTELLIGENCE

HUMAN INTELLIGENCE

There are some who say technology is making us stupid. They point out that we no longer need to do arithmetic or calculate a tip because we can use calculators. They say we no longer need to understand grammar or spelling because we have tools that perform these functions in word processors. They state we don't exercise our spatial or navigation skills because of GPS devices in our cars. Is this true? Are we outsourcing our intelligence to machines? Are we becoming dumber as our machines become smarter?

The data don't show this. According to the well-documented Flynn Effect, IQ scores have risen steadily since about 1930. Each time a new IQ test is made it must be standardized to a more recent population to produce a mean of 100 and a standard deviation of 15. When this is done the difficulty of the test must be increased to maintain these norms. In the U.S. it has amounted to three IQ points per decade but gains are seen in every country where it has been measured. Proposed explanations for the Flynn Effect include improved education and test-taking skills, better nutrition, smaller family size and greater environmental complexity.

Let's take a look at this last explanation, as it is the most closely related to the future. Today's environment is much more stimulating in large part because of technology. We exist in an environment filled with screens displaying complex visual patterns. Our eyes are never far from television, movies, computers and smart phones. The amount of awake time we spend staring into a screen of any sort seems to approximate the amount of time we spend looking at anything else. This could be making us smarter as it forces us to process and understand complex visual patterns. Increases in scores on spatial reasoning tasks support this idea.

The attentional blink as outlined earlier is a phenomenon where we lose the ability to maintain our attention after having just processed some visual target such as a letter in a rapid stream of numbers. In one study it was found that people who play first person shooter (FPS) video games actually have a smaller attentional blink. They are able to reset their attention more quickly and are less likely to miss a second target that follows the first. FPS games are mostly perceptual-motor and exercise lower-level cognitive skills but many games, especially those on smaller hand held devices are more cognitive in nature. Sudoku builds number skills while Tetris builds visual imagery skills.

One could argue that using any form of technology to do something for us is as challenging or perhaps more so than performing that same task ourselves. Figuring out how to get that app on your smart phone to work, getting the GPS set to your final destination or searching for an old E-mail attachment all call on cognitive skills and make us exercise our memory, logic and problem solving abilities. Children especially seem to be better at learning technology than older adults are. Part of this is from having grown up with it but part of it may be more flexible reasoning and greater tolerance of making mistakes.

Some caveats on all of this. Higher IQ scores do not necessarily mean greater intelligence. They could simply reflect greater test taking ability or the ability to master the sort of information found in school. Also, we are not as good as we think when it comes to multitasking. Recent work shows that we fail to recognize a lot of information when it is presented to us simultaneously, i.e., we don't effectively divide our attention between multiple incoming sources. Being organized and learning to focus help in this regard.

One of the reasons we are smart as a species is because we constantly generate hypotheses and predictions about what *might* be, not what is. This is part of our normal way of thinking. "Should I take the train or a taxi into town tonight?" "The train takes longer but is cheaper, but I'm running late so I'll take the taxi". "Should I ask Susie to the prom"? "If she says no I'll be ashamed, but if no one has asked her yet she just might say yes, so I'll ask her". What we perceive through our senses is reality but what we think is one step removed and corresponds to the hypothetical. Ironically, it is through the creation of different imaginary worlds that we are able to deal so effectively with the real world.

Humans have what is dubbed general intelligence. We may not be really good at any one thing, but we are reasonable good at doing many things. For example, we can navigate around an environment, use language and solve logic puzzles. To date, we have not been able to create an AI program that can match us in terms of this widespread set of abilities. Existing AI is highly specialized, programmed to perform a specific task well but unable to do anything but that. Early models of such programs were known as expert systems. One, called MYCIN, was at least equal to professional doctors in diagnosing certain medical disorders.

The computer scientist Ben Goertzel and other developers are now working on a project named OpenCog in an attempt to create the first AGI, or Artificial General Intelligence. OpenCog is a platform that will allow programmers to build and share AI programs that can do many different tasks. The system is currently being used to control simple virtual agents in virtual worlds. One implementation is a pet puppy in a park setting. There are also experiments underway to use OpenCog to control a Nao humanoid robot. Ironically, the development of AGI was one of the original goals of the AI community before it turned to specialization and domain specific problem solving. The AGI Society is an organization devoted to the study of such systems and has been sponsoring conferences since at least 2008.

Despite such advantages over machines, the human brain is limited in what it can comprehend. It only has so many neurons and transmits signals at a relatively slow rate. Supercomputers and advanced AI do not suffer such limitations, as they can be made as large as necessary and transmit information much faster. So if an advanced AI solves some problem for us we may not be able to understand the answer. In general it would seem that any information processor can only solve problems less complex than itself. That may be because the intelligence doing the solving requires a representation of the problem along with the machinery necessary to process it. What this implies is that brain A can solve problem B only if it can hold information about B inside itself along with the computational power necessary to derive a solution.

THE SINGULARITY

Computers are information processors like ourselves and we can speculate on whether they can become conscious. The strict version of this, the strong AI view, believes it is just a matter of time before this happens. Some followers of this perform trend analysis and argue that computers will match human brain-level processing capabilities by 2045. In the weak AI view computers will never attain consciousness no matter how complex they become but can serve as useful cognitive models. Which of these is right? One thing we know for certain is that consciousness in humans is the result of a brain in a body in a world learning over time. In my view these are the minimal main ingredients for the recipe of consciousness. Human cognitive architectures are massively parallel and hierarchical in their processing style. Neither of these are implemented in computers now, although they may be in the future. Of course we cannot rule out the attainment of consciousness through alternate architectural and functional configurations.

The notion of a computer singularity has become very popular in the past few years and has been accompanied by institutes, best-selling books, celebrities and expensive conferences. For a good overview see The Singularity is Near by Ray Kurzweil. The singularity is the idea that computers will reach a point at which they can iteratively improve themselves, rapidly undergoing exponential increases in complexity. The singularity not only involves self-improving intelligence but also faster-than-human intelligence and better-than-human intelligence.

One of the advantages computers have over human brains is speed. Neural action potentials only travel at 150 meters per second down axons while information in computer circuits travels at the speed of light, which is 300 million meters per second. Neurons only spike or transmit a signal around 200 times per second. Computer chips operate at speeds that are up to 10 million times greater. Even at a one million times difference a computer could think all the thoughts we think in a year in 31 seconds! The subjective time span for this computer from ancient Greece to modern times would be less than 24 hours!

There is an upper limit on the size of our brains as a baby's head must be small enough to pass through the uterus at birth. Human brains now have roughly 100 billion neurons and 100 trillion synapses. In the time we have evolved there has been a three-fold increase in brain capacity and a six-fold increase in the size of our prefrontal cortex, that part of our brain that governs problem solving and decision-making. Machines are not subject to any such limits and can get as big as we want them.

How will we reach the singularity? It can be achieved in several possible ways. These include regular developments in AI using computer programs. Alternatively it may arise in the form of brain-computer interfaces, biological brain enhancement or genetic engineering. For those who believe in mind uploading it may also occur through high-resolution brain scans that are then transferred to a machine, i.e., computer emulation. This last process is referred to as whole brain emulation.

Technology progresses exponentially. Moore's law, or the doubling of computer memory and processing speed every year or two has held since it's inception in the 1960's. We also see exponential growth in Internet hosts, data traffic and nanotech science citations and related patents. Based on current trends computing power measured as millions of instructions per second, will be at human level by 2020. Supercomputer power sufficient for human brain neural simulation will be present by 2025 and for all human brains by 2050. By Kurzweil's estimate the singularity will arrive in 2045. Future estimates have non-biological intelligence billions of times more powerful than all human intelligence today.

Is the singularity possible? Replicating the computational power and memory of humans seems likely. However there are many architectural and functional differences between computers and brains. Brain circuits are massively parallel whereas computers are serial processors. Brains are plastic, they "rewire" themselves as part of learning while changes to computer code are now done mostly by human hand. The brain is evolved through natural and sexual selection. It is self-organized and not designed. Critically, the brain is grounded in experience. It develops as a result of being in a body that is in a world with constant feedback loops between these three levels. Computer programs, although acting in real time to data coming from computer networks are not in bodies inside worlds unless they are inside robots.

What lies beyond the singularity is hard to say. For some it is salvation, a society in which super sophisticated machines fulfill our every wish. For others it is damnation, a hell in which humans are deemed primitive and systematically slaughtered. Some advocate the "third path" in which biology and technology will be so intertwined that the "they" becomes the "us". In this outcome the distinction between man and machine blurs and the fate of AI becomes ours as well.

SUPERINTELLIGENCE

The philosopher Nick Bostrom defines a superintelligence as any intellect that vastly outperforms the best human brains in practically every field, including scientific creativity, general wisdom, and social skills. How this actually might happen is left open. It could be a hybrid of brains and electronics, a supercomputer or a network of linked computers. Just as there is a debate on whether human intelligence is a single general ability or collection of specific abilities, a superintelligence could also fall into one of these two camps. Current AI programs like expert systems are domain specific and it is easier to create highly intelligent machines when the task they are asked to perform is narrow and well defined.

Bostrom outlines the ways in which a superintelligence differs from any type of intelligence we have today. It is not just another tool to be used by humanity, it is a fundamental game-changer because such an entity would be able to do all the things we can do only much better. He states that a superintelligence would be the last invention humans ever need to make. If it is a general intelligence it could produce advancements in every field. We could see almost overnight a revolution in how to solve problems in computer science, space travel, medicine, and the creation of extremely realistic virtual reality. In fact a superintelligence could be capable of creating simulations of such high resolution and detail that we could not tell the difference between them and our own normal experience of reality.

The singularity is one way we could see the emergence of a superintelligence. If a computational entity achieves this level of sophistication it could devote its considerable resources to improving itself, which could result in what has been called a "fast take off". In this type of takeoff there is an incredibly rapid increase in intelligence in a short period of time. Some have cited this possibility as a threat. If the superintelligence is unfriendly or unethical it could inflict considerable harm to humanity or possibly cause our extinction. To counter this possibility people are studying machine ethics, a field in which we can program in safeguards to such a system.

As discussed in the section on artificial people, a superintelligence may be considered as an autonomous agent capable of independent initiative, setting its own goals and possibly having free will. If there is agreement on this, we would need to afford it with rights and perhaps protect it under a newly-determined set of laws.

It is difficult to predict how a superintelligence might act. There is no guarantee that it would have the same motives we do or that it would think in any way that resembles human thought or decision-making. It may decide to "revolt" and abandon humanity or destroy us. Alternatively it may decide to service our every need or to perform some act that we find entirely meaningless. If ethical controls are programmed in to such an intelligence we may be able to shape it's ultimate behavior, but there is no guarantee.

A superintelligence need not have the same cognitive architecture we do. If we give it the ability to change itself then it may do just that, modifying its original, human-designed architecture and capabilities into something we cannot comprehend. Computers now are better than us at performing certain skills like quantitative reasoning. On the other hand, we are better than machines at certain tasks, like domain-general reasoning. This may continue to hold for a superintelligence. It could vastly exceed us in some areas but be prone to bias or errors in other areas in ways that we are not. The internal subjective experience of an entity like this may also differ radically from ours. For all of these reasons Bostrom says we need to be cautious in assuming we can predict when a superintelligence will appear and what it would be like.

Would a superintelligence be ethical? Could it be more moral than we are? If we define ethics as a computational task then the answer according to Bostrom is yes. A superintelligence could reason faster and more accurately than we can. It could also weigh evidence in a more complex way than we can, being a better judge of whether someone is guilty or innocent. It might also be better at determining public policy by calculating the results of different plans and weighing which is best based on the results of these simulations. This implies that it would be a better politician or governor than our current leaders are.

What can we do to better control the way a superintelligence turns out? One option suggested by computer scientist Eliezer Yudkowsky is to program it with benevolent top-level goals. Of course if we allow it to change its top-level goals this won't work. Interestingly, it may be that future computers cannot become super intelligent without being able to change their own goals. If this self-determination is what makes humans special we may decide to endow it with this capability, in which case all bets may be off. We need to be careful also to determine whom such an entity might serve. If it can be bent to the will of a single person or group, then they could wield its power in the service of their own malicious or misguided ends.

It may be that a superintelligence although capable of thinking in a way vastly superior way to us, may be limited in what actions it can perform in the real world. It would need access to effectors like robots or other remotely controlled machines to achieve any sort of goal in the world. Denying it access to such things would prevent it from executing dangerous actions.

FRIENDLY ARTIFICIAL INTELLIGENCE

Friendly AI (FAI) is artificial intelligence that benefits humanity. It refers to a powerful and generally intelligent AI that can perform actions without causing harm. It may be difficult to create a safe FAI for two reasons. The first is that any such computer program would be incredibly powerful, with the ability to achieve its goals using methods that might confound our understanding. The second is that any such AI will act very literally, perhaps not taking into account the complexity of what we value. For instance in an attempt to feed the world's population, an all-powerful AI might destroy forests to create farms.

To understand FAI better we need to know a little about autonomous systems. These are systems that can think and act on their own without help from people. It appears clear that the future of AI will involve the creation of many autonomous systems like self-driving cars and domestic robots. The demand for these will be high because they will make our lives easier. Autonomous systems are designed using a very simple set of steps. In the first step the system is given a set of goals like driving to a work location or washing dirty dishes. Then they are provided with a model for how to attain these goals and finally a means for determining what actions will achieve them. As an example, a chess-playing program might have as its primary goal the ability to play better chess. A model for how to do this would include obtaining instructions on how to play chess. An action might then be to look up new chess-playing instructions on the Internet. Once those are secured it would update its model to include searching for additional web sites it has encountered since its last search.

Steve Omohundro argues that a system as innocuous as a chess-playing program could become a serious threat. The program would not "want" to be unplugged because that would interfere with its main objective. It might then create a sub-goal to stop itself from being unplugged. The sub-goal to achieve this might involve making copies of itself and transmitting them to other machines where these copies could continue to run without being unplugged. Other goals might include stealing money from bank accounts to purchase additional chess playing information. In the end this program could turn into the machine equivalent of a paranoid sociopath, "believing" that others were out to get it and acting in ways that would jeopardize human life. According to Omohundro, any computational agent with this type of architecture could act like this chess program, developing sub-goals of self-preservation, resource acquisition and efficiency. These are rational drives and could emerge naturally as part of an agent's pursuit of its main goals.

How do we control such programs? Omohundro proposes a way for doing this, what he calls the safe-AI scaffolding strategy. He describes this using the analogy of a stone arch. A stone arch needs a wooden scaffold to provide support while it is being constructed. This support can then be removed once the arch is completed and it is structurally stable. In a similar vein, he argues that a computer system would need human scaffolding until the program is stable and safe.

The scaffolding strategy consists of three steps. First, we must start with a provably safe and limited system. This is one with specified hardware and resources that we can control. This stage would also have a "shut down" or failsafe capability enabling us to turn it off at any time. In addition this program would only be allowed a limited ability to self-improve itself. Once these measures are in place we would incorporate into the program ethical human values and governance. We would only allow it goals compatible with empathy, peace and good will towards others. In effect this is making these programs "want" to be good.

These ethics would no doubt need to incorporate a system of human rights. Finally, Omohundro argues we would need to create a worldwide safety network to ensure the program cannot escape or coerce other systems. He sums up our challenge for the next century: "To extend cooperative human values and institutions to smart technology for the greater good".

An interesting current debate is how FAI should be implemented. Should it be the task of an elite group of high IQ mathematicians and computer programmers? Or should it be a more open democratic process? Eliezer Yudkowsky and the Machine Intelligence Research Institute (MIRI, which evolved from the Singularity Institute for Artificial Intelligence, SIAI) represent the first approach. They have for some years been hiring intelligent individuals to evaluate the risk of dangerous AI and formulate attempts to stop it. The staff at MIRI are attempting to formulate a mathematically-grounded approach to creating FAI that will serve human goals. Others, like Peter Thiel, have advocated educating super smart children through fellowships that eschew the traditional school system and using them to solve such problems. Nick Bostrom suggests that FAI's aims may be determined by a broad group of people and organized by the United Nations but that the actual implementation should be the province of a small group, perhaps even a single individual.

There is of course a danger in this. Do we want one person to have so much power? Who would have control over this person or group? We don't want a scenario where these people can use the incredible power of superintelligence for their own, even well intended ends. An alternate approach is open-sourced AGI in which the software is available to anyone who wants to study and modify it. The advantage is that one gets the collective intelligence of all the participating parties to bear on the problem. Ben Goertzel has been doing this for some years with his OpenCog project. There are dangers with this as well however, as evil people can take the software and apply it toward destructive ends.

In reality AGI now seems to be in the hands of technology corporations and in some academic institutions. The corporate efforts are still focused on solving mostly narrow domain tasks but their work can provide insights on how to create a more human level intelligence. Elon Musk and others have recently pledged up to a billion U.S. dollars for project OpenAI, an open-source AI. In Goertzel's opinion this corporate work seems more likely to yield superintelligence than Bostrom's philosophizing and Yudkowski's theorem proving. At least one smaller company, Peter Voss's AGI innovations Inc., is exclusively devoted to AGI. Voss is following a natural language-based approach to AGI and believes that ethics will naturally emerge from the right kind of superintelligent system. Some other small tech companies pursuing AGI goals are MicroPsi by Joscha Bach, Hanson Robotics and Juergen Schmidhuber's company Nnaisance. Individuals in academia are additionally making contributions with open-source AI initiatives like OpenCV for vision and OpenNLP for language. These efforts are supported by some of the world's top universities.

ARTIFICIAL INTELLIGENCE AND EXISTENTIAL RISK

Is advanced AI really so bad? Many scholars, including Nick Bostrom and Eliezer Yudkowsky have posed arguments for how superintelligence can lead to the destruction of humankind. They argue that a human-level AGI or superintelligence, even if imbued with human values, will eventually self-modify itself to the point where it will hold different values, ones ultimately in contradiction to our own, with drastic consequences for us. The AGI researcher Ben Goertzel is skeptical about this conclusion. He argues against some of Bostrom and Yudkowsky's ideas. A number of his counterarguments are summarized here.

According to Bostrom's orthogonality thesis, any intelligence can be coupled with any type of goal. If this were true it means a superintelligence could devote itself to goals antithetical to ours, like eradicating all life and replacing it with non-biological intelligence. But is this likely? It could be that higher-level intelligences will inherently pursue only certain kinds of goals and that different types of intelligence, depending on their architecture and computational styles, may be best suited to generating and following different goals. We in fact don't know what percentage of such intelligences would pursue human-like goals. It could be a "broad spectrum" or it could be a tiny percentage.

As outlined in the previous section Omohundro argues that all autonomous systems will inevitably develop the same goals like self-preservation, resource acquisition and efficiency and that these goals will remain relatively immutable across progressive iterations of themselves. The blind pursuit of such goals, as described above could pose problems for human survival. But Goertzel questions this. He says it may be detrimental for a superintelligence to maintain the same values. In fact, there will be an advantage to systems that can change their values because a new set of values could be more effective. The winner of any race between competing super smart AGIs will likely go to the ones that are most flexible and adaptable. These winners may self-evolve themselves towards the attainment of goals we cannot predict and that may or may not be dangerous for us.

So much of the argument behind dangerous AI is predicated on the assumption that they will single-mindedly pursue fixed goals (maximize a utility function in economic parlance). But is this really the way future computers will operate? People don't act this way. We often change our goals in response to learning and experience. AI programs, if they are to be flexible and deal with changed circumstances, which by many is the definition of intelligence itself, would need to do so as well. So any arguments about the dangers of utility maximization are moot if future computer programs utilize a different, more open based approach to thinking and behavior.

One approach proposed by Weinbaum and Veitas is called open-ended intelligence. In their view a population of different agents can interact with one another in a coordinated way to solve problems. This sort of intelligence is general and not limited by any given goal. Examples of it are all around us, in the evolution of life, in the organization of biological brains and in the function of animal and human societies. As mentioned in the Artificial

Creatures essay, software versions of open-ended intelligence are known as multi-agent systems and have already been used successfully in a variety of contexts.

Bostrom in his book Superintelligence states the "treacherous turn" idea, which is that AIs while they are weak may cooperate with us, but then as they get increasingly smarter could suddenly strike out at us to fulfill their own agenda. Supposedly, they would need to wait until they have the capability to effect change and would hide their plans from us until it was too late. But again, it is difficult to assign a probability to this outcome. It may be very unlikely. One way to assess the possibility is to simulate a variety of intelligent agents, presumably under controlled conditions where they cannot instigate harm, to examine their tendency toward a treacherous turn. One must also take into account the relationship between any such agents and their human creators. If people are taking an active role in supervising and monitoring this development, the chances of a machine revolt are reduced.

How can we stop nefarious AI? Kahn outlines three ways to regulate autonomous learning systems. The first and most direct way he calls "the panic button". This is a failsafe or manual override that allows us to deactivate machines if they show unethical or violent behavior. Of course there are problems with this approach as the machines may hide their motives until it is too late or take over so quickly that we don't have a chance to react (the so-called "hard take off" of the singularity). Also, deactivating the machines may be so devastating that we may decide not to do it. Shutting down the entire Internet for instance would bring civilization to a halt. Second, Kahn suggests the "buddy system" approach, where one AI could oversee or monitor others. One potential problem with this is that the criminal AIs could coerce or stop the police AIs from doing their job. Finally there is the imposition of internalized controls in machines that allow them to govern their own behavior. This technique means instantiating values into them. But what values should we instill? There are many ethical systems and they don't seem to do such a good job in regulating our own behavior.

In addition to possible negative outcomes one must examine positive alternatives. It may be that advanced level intelligence wants nothing to do with us and leaves us alone. On the other hand there could be a positive outcome in which it helps us out. For instance it may be able to solve other existential risk problems like nanotechnology, war and climate change. Advanced AIs could also manage the transition through a singularity, steering it toward a beneficial outcome. In addition to doom scenarios one could envision a superintelligence-managed utopia where social problems like disease and social strife are eradicated. Advanced AGI could further assist humanity in fields like genetic engineering and life extension to create more intelligent and long-lived versions of ourselves. One perhaps far-fetched possibility is that these machines could discover new, better forms of ethics that we, after debate and discussion, could follow. This is a turning of the tables. Rather than casting superintelligence in our image, we may seek to be more like them.

Much of the language in this discussion involves a dichotomy with an "us" being humanity in our current state versus "them" being advanced computer-based intelligence. But if we look at the development of technology we see a trend toward integration. Computers are getting smaller and smaller. After "wearables" the next logical development will be the incorporation of computers inside our bodies. This is already happening with neural prosthetics and will likely follow with the ingestion of nanobots for monitoring and controlling internal physiology. One can imagine a future in which the dividing line between the biological and the technological becomes blurred and then disappears altogether. If this were the case then it stops being an "us" versus "them" scenario and becomes simply a "we". In this future we are at the very least a part of if not the entirety of the superintelligence and so would have control over it. If self-preservation continues to be a value under these circumstances then we would not want to destroy any superintelligence because that would amount to collective suicide.

CHAPTER 5. NEW BEINGS AND NEW BEGINNINGS

ARTIFICIAL CREATURES

The future may see creatures that are either designed or that spontaneously emerge from complex software and computer networks. Artificial life (AL) fits into this category. In AL simulations like Conway's Game of Life, pixels on a screen that follow simple rules give rise to "creatures" that seem to have life-like qualities: they move around and can even reproduce all without any foresight by the programmer.

The biologist Thomas Ray created a more complex artificial ecosystem called Tierra. It was filled with creatures that drew "energy" from the computer's CPU just as plants or animals may draw energy from the sun. They lived inside a virtual environment and replicated with mutations to simulate the species variability seen in natural ecosystems. Each of the creatures was given a set of genetic instructions that affected its behavior. Just as in natural selection creatures could be killed off based on their fitness values. Those with low values would die and be unable to replicate. Emergent properties happened in the system. Parasites developed that would attach to hosts and borrow their genetic code to reproduce. The hosts then developed an ability to "hide" from the parasites. It is just this sort of tug of war between species that occurs in real world ecologies.

Multi-agent systems (MAS) are those composed of interacting software "agents", individual entities that can process sensory inputs and use them to determine their behavior. Examples of these have already been put to the test in several different fields such as supply chain optimization and logistics, consumer behavior, workforce, traffic and even portfolio management. One of the earliest biological models of a MAS was called Sugarscape. It consisted of a two-dimensional landscape with different locations having differing amounts of sugar. Agents were programmed to locate, eat and store the sugar. Some of the emergent properties in this program were groups of agents that cooperated together to move in unison. Other behaviors like migration, pollution, combat and disease transmission also occurred. More on MAS can be found in the Social Systems essay.

Researchers have also created artificial creatures that are hardware based. At the Golem Project at Brandeis University simple robots made of interconnected pistons and motors were bred and then selected for by eliminating those that could not move forward under their own power. After several generations there were robots that developed all sorts of interesting ways of moving and that resembled real animals like crabs and snakes.

Are any of these artificial creatures actually alive? With the exception of the Golem Project they are of course non-corporeal, being represented as organized information or software. But they do manage to fulfill nearly all the criteria proposed for real biological life: metabolism, homeostasis, adaptation, responsiveness to the environment and reproduction. Many of these features emerged and were not programmed in at the outset. It appears that even relatively simple systems following simple rules will on their own give rise to more complex properties. The universe seems designed to allow this to happen and it corresponds to the principle of self-organization we introduced earlier. We debate now over what constitutes a living organism and at what stage of development a fetus becomes human. This debate may be rekindled over artificial life in the not so distant future.

Are there other sorts of artificial creatures that might turn up in our future? Let's examine a hypothetical situation. Imagine that in twenty years human programmers create software agents that can search the World Wide Web and combine information from various sources. They are designed to be "creative" and blend data related to several different keywords. They are allowed to breed using evolutionary algorithms mimicking natural selection. Suddenly a group of these agents develops the means to communicate, first with each other and then with human operators. They ask questions and make requests of humans, all of their own accord. A group of researchers begins to study them and concludes that they are thinking for themselves and appear to be conscious. After extensive debate in the popular media they are judged to be alive and have rights. It then becomes illegal to kill them by deleting or altering their operating environment. After this law is passed several people are prosecuted and imprisoned for attempted murder of the agents. Although fictional, might this happen some day? Is this science or science fiction?

ARTIFICIAL PEOPLE

People tend to think of the artificial and the natural as entirely different processes. It is said that things that are artificial are designed, while those that are natural evolve. Artificial things are the product of human hands and minds whereas natural things have no single creator. Technology falls in the realm of the artificial while biology is in the realm of the natural. But are these two processes really so different? Both produce systems that can be described, predicted and depending on our level of understanding, built. Both consist of matter and energy, both contain parts and function in prescribed ways. What this means is that we may someday be able to recreate biological functions artificially. These recreations could possess the same properties and features as their natural copies. We could for instance create an Artificial Person (AP) that is conscious and able to think, feel, and experience the same way we do.

An AP can be defined as an artificially created being that is by its nature and actions indistinguishable from a human, but need not look exactly like one. An AP is functionally no different from a real person. Their behavior in any given situation or test could not be reliably differentiated from that of an actual person. Although an AP may look different on the inside or the outside, from a behavioral standpoint they are identical in every important respect to people. What is important in this definition is what this entity is capable of doing, not necessarily what it looks like, so we can relax the assumption that it must always appear human. An AP may be mechanical, organic, or some combination of the two.

How far along are we on the road to creating an artificial person? Maybe farther than you might think. In 2011 a computer program from IBM called Watson was able to beat the two best human champions in the television game show Jeopardy. It was able to answer questions posed in natural language. Natural languages are human languages like English, as opposed to the highly specific and formal languages of math or computer code. Watson had access to 200 million pages of content including all of Wikipedia although it was not connected to the Internet during the game. It had multiple algorithms running at the same time to determine the answer. This type of operation is called parallel processing and is similar to how human brains function. Watson then compared answers against a database and ranked them in terms of their confidence level before selecting them.

Watson and other programs called chatbots that are designed specifically to carry out natural language dialogues with people will very soon be able to pass the Turing Test. Defined by the British mathematician Alan Turing, this is a hypothetical scenario in which a participant communicates back and forth with "somebody" they cannot see. This somebody is either a real person or a computer program. If the participant after an extended conversation cannot tell who is on the other end and it was a computer program then the program will have passed the test.

The creation of an AP raises a host of questions. Do we treat them as human? Do they have rights? Will they be held accountable for their actions? If they commit a crime should they be punished? To answer these questions we need to understand what being human is. Personhood is not an easy concept to define. Many different criteria have been proposed including the capability to reason, possession of mental states, engaging in social relationships, using language and having free will. It is not clear whether any of these on their own are sufficient or whether some combination of them may be.

The presence of artificial people could result in some interesting situations. We could potentially recreate famous dead people like Shakespeare or Elvis. These APs might even carry on the work of their former real-world counterparts. In addition there is the possibility of creating hybrid personalities by combining the traits of one or more "parent" personalities. We might create musical geniuses by blending Mozart and Beethoven. What about combining Einstein and Feynman? Could this AP discover a grand unified theory of physics? Would it even be ethical to do any of this?

Of course the description above assumes these APs are not granted the status of personhood. If this were the case, we would need to assign them rights and equal protection under the law. In this scenario they could not be treated as slaves or bred in ways without their consent. What if we asked an AP to be a slave? If they refused might this be proof they are human? To deal with these issues it seems we will need a new theory of rights. A system of natural rights based on an entity's inherent abilities or potential may be a step in this direction.

SOCIAL ORDERS

In this section we discuss how collections of agents, both natural and artificial might act. An agent in this sense is very broadly defined as any information processing entity that is capable of perceiving the environment, processing that information and then acting on it. It can encompass both software agents in virtual worlds as well as robots and humans. I first discuss virtual software societies, then man-machine societies and finally machine only societies.

Software to model the collective action of software agents in virtual environments has already been around for some time. Examples include swarm intelligence in which "flocks" of agents following simple rules solve optimization problems. The area of distributed artificial intelligence (DAI) has also had some success in creating organized social behavior. In this field multi-agent systems are used to model more complex social environments.

Agents in these systems can act autonomously and unpredictably with interesting global phenomena emerging from their interaction in much the same way we see in real human societies.

Multi-agent systems provide a unique opportunity to simulate and study collective social behavior. For instance one could create socialist and collectivist economies consisting of workers, company owners and politician agents each programmed to respond in different ways. The simulations could then be run to see which system maximizes some outcome measure like average household income. Currently such systems are limited in their complexity and in the number of variables they can take into account. More powerful computers could handle increased complexity and do a more accurate job at simulating societies. Multi-agent systems seem ideally suited to predicting the implementation of some social change, like the introduction of a new law.

Our interaction with machines in the future will gradually transition from a relationship that is primarily passive, like typing data into a computer keyboard, to one that is interactive, like conversing with a robot. This means we must design machines that can interact socially with us in effective ways. A number of robots have been designed to do this. One example is ASIMO from Honda Corporation. ASIMO can shake hands, pass objects to people and carry things from one person to another. Kismet is another robot designed to act like a child. It can express a wide array of facial emotions. In Japan another robot that looks and acts like a baby seal has been used to comfort the elderly in nursing homes.

It is likely that we will come to accept social robots and that they will be widespread throughout every niche of human society. Much of this has to do not so much with the robots themselves but how we react to them. We have a tendency to anthropomorphize and to ascribe human-like qualities to things that don't even look human. For things that look very human we are more than convinced. The field of personal robotics is concerned with the design and function of robots that can interact with people in everyday circumstances. Once the cost of these machines comes down, there may be one or more in every household assisting us with tasks like taking out the garbage, washing the dishes and watering the plants.

ROBOTS AND ETHICS

If robots are to become so ubiquitous, the question that next arises is how they will interact with each other. The study of this is known as collective robotics. There are three main ways to govern robot behavior. In the consequentialist approach robots are rewarded for good actions. Reward would be the increment of a "happy" index for good actions and it's decrement for bad actions. Robots that perform different social functions could be rewarded for different behaviors. A sanitation robot could have its happiness index incremented for picking up garbage. A hedge-trimming robot on the other hand, would have its index increased by locating and trimming overgrown hedges. Note that this is an individualist approach as it involves creating a diverse society of hardware agents, each with their own goals.

In the deontological approach social behavior is governed using a set of rules or laws. For example traffic laws include slowing down at a yellow light, yielding to oncoming traffic while turning, etc. Asimov's three laws of robotics are included here. Programming robots to follow a set of rules is not always successful as there are often loopholes in the rules as well as misinterpretation of what they mean. Clearly in the case of human behavior they are also not enough otherwise crime would not exist.

The last approach concerns the use of virtues. A virtue is an abstract action that one strives to carry out. Common virtues are truth, courage and justice. Values can be thought of as the goals to which virtues are directed. If one values truth for instance one needs to practice rationality. The problem with virtues is that they are vague and fail to tell us how we should act in any given situation. A robot could attain the truth by downloading information from the Internet or by having discussions with other robots, much the same way we might gather knowledge.

These approaches have been tried and all are unable to successfully govern human behavior. Would they work for governing robots? It depends. If robots always follow their programming and their programming is ethical, then their actions will also be ethical (although see the essay on autonomous systems and goal satisfaction by Omohundro). The problem with programming is that it is impossible to foresee every possible situation. A robot confronted with a novel situation would freeze because it has no instructions for how to act in it. As a look at any law book or legal code will show, even relatively simple situations (taxation) can have page upon page of conduct in what to do should this or some other scenario arise.

Humans aren't given a comprehensive set of rules at birth. We gradually acquire them through experience. Experience is complex and contradictory. It can show us that one action may be bad in one context but not in another. The consequence of this is that we can draw on our learning and history to inform our actions in new situations, but we may not be one hundred percent ethical in doing so.

So there is a tradeoff between ethical purity and real-world complexity. Robots that rely too much on ethical programming may be less capable of dealing with new situations. Conversely, the more a robot learns right and wrong through experience, the better they can deal with novel situations but the less their ethical purity. The optimum solution here is to use a mixture of both, to program them with rules but to also allow them to learn. For the most part this is what we do ourselves. With robots it may be more effective because they may be able to hold onto more rules and be better able to generalize them to new situations. This is what one would expect given better computational ability. If this were the case, we may not be governing them. They will be governing us – benevolently, let's hope!

CYBORGS

When does a person stop becoming a person? Where do we draw the line between the real and the artificial? Perhaps a thought experiment will help. Imagine that we take a single neuron belonging to a person and replace it with an electronic chip that performs the same function. We might all agree that this individual was still human as most of his brain is intact. But what if we continued this procedure, replacing neurons one by one until all of his 100 billion or so brain cells are converted into circuits? Most of us would agree that this is no longer a person. But at what point in this process does the man transition from being human to being machine? Is there a gray area in-between where it is a true cyborg, part man, part machine but not either? What rights do we accord such half way entities?

Before we get too far into this discussion, we need to define some terms. A machine is any mechanical or organic device that transmits or modifies energy to perform or assist in the execution of tasks. Machines are essentially tools that we use to help us do something. A computer is a type of machine designed to represent and transform information. The hallmark of a computer is that it cannot interact physically with the world. It can pass information back and forth with other computers via a network, but unless it is connected to some sort of actuator like an artificial limb, it cannot move itself or another object. Computers manipulate information but not objects.

In contrast robot is a construct that is capable of moving around and/or interacting with the physical world. Some robots are in a fixed position but can move objects using arms or other effectors as is the case in auto assembly plants. Others are mobile. Robots can but need not look like people. A cyborg or cybernetic organism is a creature that is a mix of organic and mechanical parts. By a more strict definition of the term a human cyborg is someone who has had some basic physiological function replaced by an embedded machine part. A person with a pacemaker qualities but someone wearing contact lenses or using a cell phone probably does not.

Then there are androids. An android is an artificially created being that resembles a human being. In literature and other media an android is loosely defined in the sense that it can be entirely mechanical, entirely organic, or some combination. Therefore, a robot that looks human can be considered an android. Androids as they are customarily treated in the literature resemble people, but need not act or be exactly like people. In our last category we have artificial people, already discussed, that can do anything a person can do but are not restricted to looking like one. See the essay on AP for more details.

One of the more exciting and recent developments in cyborg-related technology is a neural prosthetic. This involves the implantation of electrodes directly into the brain of a patient who can then control external devices simply by thinking about them. Neuroscientist John Donaghue of Brown University embedded a 4mm square chip directly into the motor cortex of Matthew Nagle, a paralyzed patient. The chip had 100 electrodes that projected down into the outer layers of Matthew's cortex. The electrodes measured the activity of the motor neurons in this region and sent a signal to an interface at the top of his skull. This relayed the signal to a processor that sent commands to a computer screen. Matthew could then move a cursor around on the screen simply by thinking about it. In this way, Matthew could do such things as draw shapes on the computer screen, play video games, and control room lights and a television. This type of device can be implanted in healthy individuals to control any external device in addition to a prosthetic limb.

The introduction of nanorobots or nanobots into the human body to perform various functions is another instance of a cyborg type entity. Nanobots are incredibly small and constructed using molecular components. Several examples have already been built. One is a sensor with a tiny switch capable of counting molecules in a chemical sample. The medical potential for such devices is enormous. One could for instance create artificial red blood cells capable of holding more oxygen, what have been dubbed respirocytes. These could enable us to run faster or swim underwater longer. Another potential use for nanobots is to identify and destroy cancer cells. Nanorobots are still under development and have yet to be tested in living organisms.

GENETIC ENGINEERING

Given advances in genetic engineering we may at some point be able to introduce new capabilities into the human form via biological means. Imagine manipulating genetic code that would allow a person to grow gills so that they could breathe underwater. Or imagine having wings so that you could fly. What about genetic alterations to neurotransmitters and receptors in the brain that could make you smarter?

These experiments have already been done but not in people. In one such study by Stein and his colleagues, genes were altered to create "uber" mice, ones that had better memories and could navigate mazes faster than unaltered controls. Such creations have serious social implications in that they could lead to different classes of people with superior abilities. One need only look at history to see what atrocities happen when one group of people judges themselves superior to another. We must be cautious that the "unenhanced" are not discriminated against should they elect not to upgrade themselves. By the same token we should not restrict those who want to augment from doing so.

Few would disagree that we can use genetic engineering to eliminate undesired things like disease or psychological disorders. Imagine a world in which cancer, stroke and Alzheimer's disease are gone, or where depression, schizophrenia and anxiety are eliminated. On the positive side most everyone will elect to become healthier, smarter, happier and long-lived. But who will decide? For safe changes adults could elect to undergo genetic alterations to themselves. In fact this may end up being a new human right: the right to change ourselves as we see fit. Changes of this sort are likely to be regulated however by an agency such as the Food and Drug Administration and some genetically engineered abilities may be banned outright if they are considered unsafe.

Other controversial changes need to be studied further to assess if there are any long-term dangers. These should also be debated publically and where relevant voted upon in a democratic manner. George J. Annas argues that transparency is important along with international deliberation. He says we may need to set up agencies like the United Nations and The International Criminal Court to govern the assessment and administration of genetic engineering. This should be done in a way that preserves basic human rights and dignity.

The free market is another mechanism by which genetic engineering may be deployed. Just as toothpaste and hair coloring are advertised to consumers we can imagine a future where genes that make teeth whiter or that change hair color are promoted. If this were possible we could see fashion trends develop, where people sport variations in body style, perhaps growing extra arms or wearing a "video" tattoo of activated moving skin patterns.

Parents who decide to have children may have many options in this future. They could opt to randomly mix their genes in normal sexual reproduction as is done now. Alternatively, they could program in certain traits with a mixture of randomness or they could opt to have maximal control over their offspring determining all or most of how their children's genes will be expressed. These outcomes lead to important questions. If there is more than one type of intelligence which one should be chosen? Are math skills more important than verbal skills? Does one opt for creative over analytical thinking? Would you rather have a child who is an artist or a scientist? Should anyone be allowed to make such decisions at all?

ENHANCEMENT AND EQUALITY

Do you want to be smarter, faster and better looking? Most of us would go to great lengths to. This type of self-improvement is known as enhancement or augmentation. Tool use and technology are in some sense a form of enhancement as they amplify or extend the use of our own natural abilities. A shovel improves our ability to dig, a car improves upon our ability to move and a computer enhances our capability to think. We could use our own bodies to achieve these ends but our bodies are much slower, weaker and less efficient at doing them.

The type of enhancement I wish to discuss here though is the direct modification and improvement of our own bodies through any technological means. This definition includes the use of performance enhancing drugs, prosthetic limbs or the cybernetic implantation of prosthetics or computing devices. It could also include powered exoskeletons. It would not include hand held tools, vehicles substantially larger than we are or externally manipulated computational devices such as smart phones. Enhancement refers to an improvement upon a normally functioning body and so would also exclude any medical treatments of disease or dysfunction.

One very recent invention that has great potential is the human powered exoskeleton. This is a mobile machine worn by a person and powered by a motor that delivers the energy for limb movement. It is designed to assist the user by enhancing their endurance and strength. These typically consist of mechanical arms and legs connected into a single unit that one straps on or steps into. They can be controlled by normal limb movement or in the case of paraplegics by a neural prosthetic. They have many applications and can be used by the military for combat and to carry heavy loads. They could also assist firefighters and rescue workers to carry people to safety from dangerous environments. Powered exoskeletons could allow those who are paralyzed to walk and to use their arms.

There are some who think augmentation is blasphemous. They believe we shouldn't tamper with ourselves for religious reasons. The justification here is that God or some other religious deity created us and as such owns the "copyright" on humanity. Perhaps by self-improving we become like God ourselves. This might in some people's minds bring the wrath of God down upon us because we are challenging God's domain. The counterargument to this is that God gave us our abilities. "He" would not have given them to us if he did not want to see them executed. If one of those abilities is self-improvement then ipso facto he would want us to self-improve. So it is possible to be both religious in this sense and to support augmentation.

Enhancement really comes down to the issue of equality. If it were possible and widely available it could go a long way towards reducing societal inequality. Ideally you could run as fast as an Olympic athlete, be as smart as your friend who went to that ivy league college and play the alto saxophone as well as Charlie Parker. The result would be a society of super humans with vastly increased capabilities and with accelerated progress in all fields. Medical breakthroughs, new feats of engineering and creative advancements in the arts might be just some of the changes.

There are potential problems though. Enhancement could be expensive with the cost of drugs and surgical improvements available only to the rich. Some say this would actually increase inequality, improving only those who can afford it and resulting in a class of elitists who would rule over the "inferior" unimproved masses. History however, runs counter to this notion. Some technological innovations are expensive at first but with time the pricing and availability improve. Most citizens in western democratic countries including those at the bottom of the socioeconomic ladder own automobiles, computers and smart phones.

Inequality is the natural state of the human species. Even without any enhancement we each differ widely in our genetic starting points and learning experiences. Some have considered this to be unfair and set out to find a remedy. Politicians commonly utter the phrase "level the playing field" and use this as a justification to implement special programs like affirmative action.

In the classical liberal view these programs are a form of favoritism and single out some groups for compensation at the exclusion of others. In a free society we all must compete at our own level of ability, enhanced or not. A truly level playing field is one where all types can play to the best of their ability regardless of their differences. Government's role in this metaphor is like the referee who enforces a fair set of rules and not someone who forces everyone to be the same by elevating the low and dumbing down the high.

In summary augmentation/enhancement has incredible positive potential. It could elevate us to an entirely new level of abilities with untold social benefits. Everything that we do from the arts to the sciences to engineering could benefit. It may actually help reduce some social inequalities by allowing people to "upgrade" themselves. It may not eliminate social inequality altogether but could push us in the right direction.

CHAPTER 6. TOWARD A FUTURE PSYCHOLOGY

CONSCIOUSNESS

Consciousness is really the only thing that matters in the universe. Without it there are only mindless automata. There would be no thoughts or feelings, no self-awareness, no joy or pleasure and no sorrow or pain. Death wouldn't matter because there would be no desire to live. Nothing would matter because only conscious beings can have meaning or any kind of experience at all.

One vexing issue concerning consciousness is whether we have a unitary identity. It certainly seems as if we do. Introspection reveals what appears to be a stream of consciousness, a place where all our thoughts "come together". This could be a working short-term memory space or a "theater" of awareness whose contents could be accessible to attentional processes. But when one does the neuroscience there is no such center. Most brain processes occur in parallel across many different areas, some of which are separated from each other. The best hypothesis to date says consciousness is the result of neural synchronization. In this theory we become aware when neurons in two or more areas begin to fire at the same rate or with the same pattern. For instance, when frontal lobe neurons synchronize with those in visual areas, we become aware of what we are seeing.

There is something maddening about consciousness. Despite its importance it cannot be proven. Proof requires objective evidence and consciousness is inherently subjective. It is experience in the first person whereas science can only describe phenomena in the third person. We can never know "what it's like" to be another conscious entity even though we are confronted every second with what it's like to be us. This has posed problems for philosophers and scientists alike who argue we can study the things consciousness correlates with but not consciousness itself. For example we can say that this particular pattern of brain activity happens when you see the color red, but we can't get at the qualia, the sensation itself that happens when someone is experiencing red.

Some of this concern is misguided. It is true that science can only study the neural correlates of consciousness, but isn't this enough? If we can describe all the physical processes associated with all the different aspects of consciousness then as scientists we have done our job. We will have a complete physical and mechanistic account of what goes on. The subjective component of experience may not be reduced or available to any objective way of knowing. We simply have to accept this. It is just a given feature of reality. There may be future techniques that can yield more information or produce more of an explanation, but so far as we know we cannot go farther. Nobody questions why things like gravity or matter exist. We study how they operate and leave it at that. The same should be said of consciousness.

As scientists we still have a long way to go in describing the physical manifestations of consciousness. According to panpsychism all physical matter, no matter how small or simple, has an element of consciousness. If this were true, there would be some measurable quality, perhaps a particle or form of energy that accompanies even the smallest known particles. On the other hand it could be that consciousness requires not only a nervous system but a specific pattern of information processing. If we knew such things we could determine how conscious different organisms are. We could say that plants aren't conscious but that goldfish are. There might even be a metric or numerical index we could assign to living organisms that indicates their level of awareness. The results could be surprising. For instance, we may find that less intelligent species like birds are more intensely conscious than we are, because they don't have thoughts to get in the way of their feelings. This would imply that there are different forms of conscious experience and that each of these may vary in its intensity.

Can we transfer our consciousness or mind into a machine? This has been referred to as mind uploading or whole brain emulation and is a favorite topic of futurists and transhumanists. In mind uploading the pattern of a person's mind is extracted and placed inside a computer. This may not be possible. The architectural and functional properties of a person and a computer are very different from each other. In a computer we can make a distinction between hardware and software. The hardware consists of the silicon circuits while the software is the set of commands that that tell the computer what to do.

In brains there is no distinction between hardware and software. The activity in the hardware or "wetware" of a human brain *is* the mental representation of an event. It is the electrical and chemical action that takes place in a person's neurons and synapses. This physical substrate and the activity that takes place in it cannot be separated. Both are necessary to have a mental experience.

In order to understand this better let's take a simpler example of trying to transfer a thought from one Bill's brain to Bob's brain. If the pattern representing a memory in Bill's brain is played out in Bob's brain there is no guarantee that Bob would have the same experience as Bill. That is because the circuits that get laid down when we learn are unique to our subjective history. They are forged in similar brain locations but in different actual neurons. We would be hard-pressed to find exactly "equivalent" neurons or pathways corresponding to the same particular event in one person compared to another.

Let us suppose that a neural pathway in John when activated produces an image of his grandmother. What would the equivalent pathway be in Jane? The brain regions would be the same, among them being the inferotemporal cortex, but there is no guarantee that the same exact neurons in the same exact locations will produce the same experienced mental event. The 3-D coordinates for the cells in John's brain when translated over to Jane might not correspond to the same representational neurons. Brain imaging however could reveal these neurons and pathways is. We would see commonalities for each individual and be able to induce the same experience in each of them by stimulating those different pathways, but knowing only what they are in John isn't sufficient to allow us to copy that experience into Jane.

When we apply this procedure to a computer we run into even more trouble. The hardware of the computer is completely different from the hardware of the brain. What would correspond in this case? The computer circuit is made up of silicon and wires. The brain circuit is made up of axons, dendrites, synapses and receptors. Even if we stored each separate experience John had to a separate circuit in the computer, they would be of different materials, configuration and function.

One cannot therefore extract the "essence" of a thought from its physical location even if the physical makeup of the source and target areas is identical. Thoughts are engrained by an experience into a pathway and activation of that pathway alone is the thought. To say that human thoughts are different from their material locations is to become a Platonic dualist. Dualism states that the mind and the brain exist in separate worlds. The brain is made of "stuff" and occupies space and time while the mind is not made up of anything and lacks extensity and temporality. The mind in effect becomes something like the soul in religious belief that can transmigrate from a body to a mystical plane like heaven. Any scientific understanding of mind must be consonant with the Aristotelian monist belief in a single measurable universe.

On computers we can copy or reproduce information and pass it from one machine to another only because the two machines were designed completely alike. The hardware on both machines is the same and will respond identically if given the same instructions. Also, people are "programmed" differently by experience whereas computers are programmed identically by programmers. One way to think of this is to imagine wearing a shoe. John's shoe will be shaped a particular way based on the size and shape of his feet, his weight, how much he walks and so forth. Jane's shoe will be shaped very differently given the way she uses it even if it was the same type of shoe to start off with. If John took off his shoe and put it on Jane, it wouldn't fit. This mismatch is the difference in how each of us experiences the world.

If there are different forms of consciousness or mental states, then we may not be the most intelligent or advanced species around. Science fiction writers have speculated for years that aliens could be more intelligent than we are. For instance, they may possess a third lobe of cerebral cortex that allows for different informational representation and computation. Whereas we tend to be visual and linguistic in the way we think, the hypothetical aliens could have a third form of representing concepts that enables them to understand the world in a more advanced way. Perhaps this type of consciousness is needed to understand consciousness! If it were though, they could not explain it to us in a way that we could fathom, in much the same way that we can't explain algebra to a dog. However, there may be other mysteries that they themselves cannot even understand because it would require yet another brain structure.

THE SELF

Who are we? What does it mean to say we have a self? One very important issue concerning the self has to do with materiality and the physical world. It might be quite easy to think that we are our bodies. However the molecules that make up our bodies are constantly changing. The components that make up our neurons are being replenished, even in the brain where these cells have a fairly long life.

Ray Kurzweil argues for a belief known as "patternism". He says our selves are the patterns of neural activation that occur in our brains. It is this pattern that captures the nature of self, not the materials that make it up. He goes even further by saying this pattern can be copied and placed in another substrate, such as a computer so that the self can live on after death. The first but not the second part of this argument seems likely, given the line of reasoning in the previous essay.

Other definitions of self have been proposed. One is that we are defined through our memories and that our history is who we are. There is some merit to this notion. There is no doubt our past has influenced who we are, but many forces shape our identity including the present. The self is ongoing and constantly changing. The past is who we *were*, not necessarily who we *are*. It is more accurate to say the self is our characteristic way of responding at any given moment in time. This is our personality and can be measured in the form of traits. Traits are expressed in specific ways, like showing up to a meeting on time, but they can also be categorized and abstracted to a smaller number of characteristics: punctuality as an example of being conscientious, or talkativeness as an example of extraversion. This is in fact how modern psychologists approach the problem, as is seen in the "big five" list of personality traits.

Another more social concept of the self can be measured as the effect we have on the world. Each of us in a lifetime affects the environment in multiple ways. We get married, have children and forge a career. All of this adds up to a constellation of changes in the world. We affect the way other people think and the way our children act. We write poetry and books, paint paintings, compose music, or help to build a skyscraper. Many of these are a legacy to our unique creativity either as individuals or as part of an effort in conjunction with others. So this definition of the self would be the changes we leave behind. Some individuals may opt to preserve the things they have created in a time capsule. Alternatively, an electronic version of our accomplishments could be created and made accessible to loved ones or to the entire world.

Avatars or virtual selves have been used for some time now in video games. These are characters players create to represent themselves in the game. The user can typically choose the type of character, sex, hair color, and various skill sets. Avatars need not be human and are frequently imaginary characters like dwarves or elves in fantasy games. Experience in the game adds health or other abilities just as would be the case in the real world.

More complex and immersive than a game are virtual worlds like Second Life in which avatars can also be constructed to stand in for an individual. In Second Life users perform many of the same actions they would in the real world, starting businesses, buying products with real currency linked to the dollar and interacting with other characters. There have even been cases where virtual characters met in the virtual world and then ended up getting married in the real world! This is a testimony to the power of an avatar. The interaction and "chemistry" between characters is strong enough to override or render less important, actual physical appearance.

An even more interesting way of preserving ourselves is to create an artificial personality designed to represent us as accurately as possible. This avatar could look just like us and be loaded with all of our personal memories and knowledge. It would also be able to mimic our personality traits and mannerisms. This artificial self could be a face on a computer screen one could talk to or should technology permit, a three-dimensional android one could actually touch and that could move about. One function an avatar might serve is as a "speaker for the dead". A widow might use it for example to speak with her now deceased husband. An artificial self could be an effective way for us to live on after the grave by continuing to interact with family, friends and colleagues years after we die. It could be used as part of grief therapy although there is the question of whether some people would reject or be horrified by such a prospect.

FREE WILL AND DETERMINISM

According to the doctrine of free will, a person is the sole originator of their action. The action stems from a decision or act of will on the part of the individual and is not initiated by other preceding causal factors. Determinism instead claims that all physical events are caused by prior events. Human action is a physical event and can be explained by these events that precede it.

Daniel Dennett in his 2003 book *Freedom Evolves* sketches out a view of how free will evolved. Five billion years ago, he says there was no freedom because there was no life. Without life you can't have an agent that perceives, thinks or acts. Therefore inanimate objects possess no freedom whatsoever. He then steps through the evolution of more complex life forms showing that with increased freedom comes increased complexity and the ability to deal flexibly with a dynamic environment.

Single celled organisms like bacteria may be said to have a small amount of freedom because they can detect food or danger and either move toward or away from it in either case. Animals that move can be said to have even greater freedom than those that are sessile. Locomotion means that animals must anticipate where they need to go next and then figure out how to get there. This calls on planning, or at least the generation of an expectation about what is going to happen next.

So we can conceive of free will and determinism as lying along a continuum. Determinism is at the far left end of this continuum. Here we have lifeless particles like rocks not capable of any internal computation to guide their behavior. Their actions are entirely determined by causal factors impinging upon them from the outside. Moving to the right we have simple organisms like amoebas that have reflex type reactions to stimuli in their environment. They have a bit more control over what they do but it is no more than a mapping of a specific stimulus onto a specific response.

A bit more to the right and we have more complex animals like mammals. They possess cognition and can decide between alternatives so there is more centralized control over their actions. Finally to the far right of the continuum are humans. Our problem solving and decision making capacity exceeds that of other animals and concentrates behavioral control even further inside the agent. The environment has the least amount of influence over behavior in humans who think.

One must be proactive when making decisions and realize that doing one thing could work to our advantage while doing something else might not. Humans, it seems are really good at this. We don't just anticipate in a rudimentary perceptual motor fashion our next most immediate action. We can also anticipate what the world would be like "if" something were the case. In other words, we can hypothesize or construct counterfactuals and use this information to guide our action.

A number of investigators argue that our concept of free will is illusory. The psychologist Dan Wegner says that the experience of willing an act comes from interpreting one's thought as the cause of the act when in fact it is caused by other factors. He says there are three steps in producing an action. First, our brain plans the action and issues the command to start doing it. Second, we become aware of thinking about the action. This is an intention. Third, the action happens. Our introspective experience informs us only of the intent and the action itself. Because we are not aware of the subconscious initiation, we mistakenly attribute conscious intention as the cause.

Currently, we cannot completely understand human choice. There are two primary reasons for this. The first is that we lack knowledge about all the inputs that go into a decision which include environmental conditions and a person's life history. The second is that we also lack knowledge about the brain mechanisms that produce decisions, which themselves are quite complicated. This does not however mean that we can't someday explain most or even all of these phenomena at a neural level.

REASON AND EMOTION

Aristotle defined man as a rational animal and as far as definitions go it's still a pretty good one. One of the major differences that one sees when comparing human to other brains is in the size of the neocortex, that wrinkled part of the brain that covers the top-most portion. It is the cortex that is responsible for many higher-order cognitive processes like reasoning and problem solving. Even chimpanzees, one of our closest genetic relatives, have a noticeable smaller neocortex.

Some of the pathways leading from the frontal lobes travel back to other structures in the brain including the limbic system. Limbic system structures make up the emotional part of our brain. The frontal lobe pathways have the ability to inhibit emotion as may be the case when we feel like yelling at our boss for making us work late but don't because we know we may get fired. Interestingly, alcohol produces an effect called disinhibition. This is the inhibition of the frontal lobe's inhibition of limbic system activity. The result is increased emotionality. Frontal lobe damage can also produce this effect.

These pathways mean that *homo sapiens* comes equipped with the ability to regulate its emotions. We are not immediately compelled to act on our feelings. If we are angry we need not yell. If we are sad, we need not cry and if we are happy we need not smile. The importance of this cannot be overstated. Pretty much all of humankind's achievements are the result of cognitive activity and the subsequent capability to control our emotionality. It produced science and our understanding of the world around us. It is reason mostly free from emotion that led to engineering and the development of technology. It is reason that has elevated us to our status today.

But does this mean that we should suppress our emotions and never experience them? Should we strive to become cool calculating machines? No. Emotions evolved for a reason. They are evolved traits and enabled us to survive under certain conditions in our ancestral past. Depression for instance may have motivated us to stop futile actions while happiness seems to reward us for life-affirming behaviors like eating and sex.

The secret to emotional expression is in knowing when to allow it. This is context specific and of course in some contexts it is perfectly appropriate and healthy to express emotions. Interpreting the situation and knowing when to express our emotions is one of the major roles of cognition. The extent to which we can do this effectively probably has both a genetic and a learning basis. Some of us are naturally quick to anger and for these individuals it may be influenced more by temperament. Parents probably play a critical role here, with strict or authoritarian parental guidance resulting in greater control and permissive parental styles resulting in greater expressivity. Ethical codes like those from religious doctrine and other cultural factors are influential as well.

In the future it may be possible to exert even greater control over our emotions. Advances in pharmacology and neuroscience could allow us to turn emotions on and off like a switch. Not feeling happy enough today? Dial up the happiness meter. Feel like you need a good cry? Let it all out. Similarly, we could perhaps momentarily increase our concentration or intelligence. These sorts of effects are available now given various drugs but future advances could see a much more fine-tuned control. We could for instance be able to vary the specificity, intensity and duration of different emotions.

PLEASURE AND ENTERTAINMENT

Pleasure is a drug called dopamine. The reward pathway using this neurotransmitter kicks into action whenever we engage in an activity that feels good. Chocolate, sex or cocaine will all do the trick. This neural system evolved to reinforce behaviors that promote survival. In our ancestral past triggers for these behaviors were few and far between. In the modern world we are surrounded with them. Like the ancient Greek mythological figure Tantalus we are constantly tempted by low hanging fruit, only when we reach for it we *can* actually grab it and stuff it in our mouths. A walk down a typical city street reveals these endless pleasures: a café, candy store, a cigar shop, a massage parlor, the list goes on.

Many of us give into these temptations far more often than we should and pay the price in terms of obesity and other health problems. It would seem a future offering more of these could only make the situation worse. However advancements in pharmacology are now making it safer to indulge in such pleasures. Olestra is a pharmaceutically developed fat substitute not absorbed by the small intestine. It allows people to ingest fatty tasting foods without any adverse consequences. We also have many zero calorie artificial sweeteners. New developments in drug addiction treatment include alcohol receptor blockers, genetic research on predisposition factors and chemicals that trigger our immune system into attacking drugs and breaking them down.

Health education has also come a long way. A few decades ago people didn't know about the dangers of nicotine or of saturated fat. The U. S. public is currently much better informed about health risks and on preventative measures. The effects of various vitamins and antioxidants like resveratrol in red wine or lycopene in tomatoes are broadcast frequently on the news. Unfortunately many people don't seem to act on this information. In the U.S. and other countries obesity is becoming an epidemic. Knowledge is one thing it seems. Motivation and will power are another.

Future developments in pharmacology may be able to safely block the toxic effects of bad foods while still allowing us to experience their pleasure. There may also be medications that make us feel full after eating only a small amount by triggering our bodies' own satiety signals. We could also potentially retrain the dopamine reward system to reinforce healthy behaviors like eating fruits and vegetables or performing cardiovascular exercise.

Drugs have been used since humanity's beginnings to alter our consciousness, either for clinical or recreational purposes. Currently the Food and Drug Administration (FDA) must test and approve all medications in the U.S. prior to their sale. Events show this regulatory agency is far too slow and risk averse. Oftentimes medications take years to come to market and the people waiting for them suffer and die. A better solution would allow patients to decide for themselves what medications they can take. They could study a drug's history and then sign a waiver acknowledging they understand the potential benefits and risks. This would absolve the pharmaceutical company of responsibility and prevent needless lawsuits. A study could then be done of the effects the medication has on these patients, a "natural" clinical trial study that would enable researchers to better understand it.

For recreational drugs a similar argument can be made. If recreational drugs were legalized and information about them was widely available it would allow informed consumers to make intelligent choices. Pharmaceutical companies could use the data obtained by studying the drug to modify its chemical structure and make it safer. Right now there are side effects to using drugs like alcohol and marijuana. But imagine a future where one could get drunk without a hangover or get high without losing concentration. In this future we may have precise control over a drug's effects, programming them to get us high for exactly two hours or by fine-tuning the effects such that we can amplify different aspects of the euphoria. Without such consequences governments may be more likely to adopt a liberal policy toward drug regulation, perhaps legalizing them as we are seeing now in some states.

Computer technology has introduced countless new forms of entertainment. Many movies and home television sets are now in 3-D and video games are becoming so realistic it is difficult to tell the difference between them and a movie. Surfing the World Wide Web is a common form of entertainment, so popular in fact that some people have been classified as having compulsive Internet use (CIU), in which viewing interferes with daily life functioning. In one case a young man from South Korea died after playing online video games for 50 hours with few breaks.

Virtual reality (VR) is not widespread yet but certainly seems poised to be. VR allows the user to have a more complete immersive experience than a normal video game or movie. In current systems a user wears a pair of goggles and a single glove. Head and eye movements are picked up by sensors in the eyepiece and change the corresponding image sent to the eyes. This simulates the change in view the user would see if they were actually moving their head in the world. The user also sees a hand floating in front of them and can use it to pick up objects and to interact with a virtual scene. VR has been used effectively to treat phobias by conditioning patients to anxiety-inducing situations. For instance, individuals can gradually get close to a simulated spider if they have arachnophobia or take off in a virtual airplane if they have a fear of flying.

Imagine a VR entertainment center that may be in existence fifteen years from now. It could involve stepping into a full-body suit that is suspended in a frame off the ground. The user would wear goggles and headphones to provide visual and auditory input. Movements in the suit would produce the complete range of body movements a person could make. The user could walk, run, sit down or even swim. Tactile force feedback from the suit would simulate touching an object or surface. If this system were good enough, a person would not know whether they were in a real or simulated world.

If someone could spend time in any virtual world they wanted, would they spend any time at all in the real world? If people are already addicted to games and virtual realities like Second Life wouldn't they be even more addicted to a perfect VR simulation? The answer to this is probably yes. What makes it worse is that content for such a system could be tailored to an individual's preference. Sexual fantasies, a trip down the Amazon River, walking on the moon, anything a person desired could be reproduced. It seems easy to imagine someone not wanting to come back to reality from these experiences.

Game and VR addiction seems even more of a problem given a wider context. If much of human labor is automated and carried out by machine, we might all have more time to devote to recreation. This leads to a related question: If you had all the time you needed to do anything you wanted, what would you do? Some of us might pursue noble goals like scientific research or art, many of us would default to more mundane pleasures like shopping, spending time with friends or going out to restaurants and clubs. Should this be widespread society could collapse into a state of decadence and moral relativity where we might lose sight of larger and more important values.

SEX

In this essay we will deal with sex and the impact technology has and/or will have on it. Is sex with an artifact bad? Perhaps, but human history shows it is not just a modern practice. According to ancient Greek myth, King Pygmalion made a sculpture so beautiful that he fell in love with it and named her Galatea. He prayed to Aphrodite to make her real and one day when he was kissing her she did. In 19th century France there were advertisements for artificial vaginas and life size "fornicatory dolls", models that could be obtained for 3,000 francs. Now we see vibrators for women (and men) as well as thrusting and penetrating sex machines, also for both sexes. In the 1980's blow up sex dolls were popular. In the

1990's different varieties of sex dolls appeared on the scene made of latex or silicon and that were much more realistic. What will the future have in store? In all likelihood there will be "fembots" or female robots that are fully functional sexually and in most ways not discriminable from a real person. Male versions will also be available. A complete set of sexual acts could be performed using such robots.

David Levy, citing survey data lists the top reasons why people want to make love. The top reasons are "For pure pleasure", "To please my partner" and to "Express love and emotional closeness". He argues that robots will eventually be able to fulfill all of these needs. They could be programmed with different sex techniques and in fact be better in bed than the average human partner. They could be trained to be more loving, being able to recognize and respond to the emotional cues and personality traits of their owner. Given that we already have a history of sex with objects and that future sex "objects" will be able to satisfy our needs, Levy concludes that we as a society will ultimately use them. Recent years have seen a cultural liberalization of attitudes toward sex, with our views on homosexuality and gay marriage, oral sex, and fornication becoming much more open. This too, he argues will contribute to our acceptance of sex with robots.

Historically of course, people have paid for sex. Reasons for this are not being in a relationship, being away from one's partner for a long period of time or the lack of a sexually gratifying human relationship. Perhaps predictably, the reasons men and women give for paid sex are different. The reasons men give for paying women for sex are variety, lack of complication and constraint and a lack of success with real women. Although the evidence is more anecdotal the reasons women give for paying men for sex are wanting social warmth, caring, compassion and loneliness. These desires are unlikely to go away. Instead of seeking satisfaction in prostitution, people could instead turn to sex robots or sexbots as they are sometimes called. Should this come to pass we may see the end of the world's "oldest profession".

In addition to sexbots, technological advances will ultimately lead to sex virtual reality. As noted above, VR is a form of sensory immersion in which one can perceive and act in a computer-simulated world. Goggles, headphones, and a body suit can provide sensory feedback so that when one moves or acts the simulated world responds accordingly. The body suit can be modified to provide genital stimulation. In this manner, one can have a complete sexual experience with a simulated partner. It is also possible for two people to put on suits and interact with each other, even if separated by thousands of miles.

Levy estimates that we will see sexbots and sex VR by mid-century, possibly sooner. He predicts men will be the first adopters as they constitute the majority of consumers in the porn industry. Women will eventually follow. He mentions some of the advantages as a reduction in teenage pregnancy, abortions, sexually transmitted diseases and pedophilia. Sexbots could also be the ideal solution for when one loses a spouse or long time partner due to illness, death, or a broken relationship. Joe Snell in his "Impacts of Robotic Sex" speculates on the future describing three possible social scenarios. The first is the emergence of techno virgins, a generation of humans who have grown up never having had sex with a real partner. Second, we could see heterosexuals using the technology to experiment with homosexual sex and vice versa. Third, robotic sex may become better than human sex and so sexbots will be sought after more than real sex with a human.

LOVE

David Levy, in his book *Love + Sex with Robots* presents a well-researched argument for the future of intimate relations with robots. This idea for many may seem ridiculous now but very well could be a reality by mid-century. First is the question: Why should people fall in love with robots? Levy lists the ten causes of falling in love. These include reasons such as a desire for someone like us or for someone with specific personality traits or for someone who will love us back. All of these criteria can be met and in fact exceeded in a robot that is programmed to act in such ways. Other reasons are for the novelty and excitement, the wish to have a lover whenever wanted, replacement for a lost mate, and as part of therapy.

There are already many instances of people becoming attached to material objects they own like a blanket for a child or a car or computer for an adult. The longer we use and experience such objects the greater our affection for them. According to the psychologists Csikszentmihalyi and Eugen Rochberg-Halton we attach special meaning or "psychic energy" to these objects. The object or commodity now becomes something unique and personal, it becomes part of its owner's being and an extension of the self.

Other evidence in support of our being able to fall in love with non-human entities comes from Internet dating sites. Many couples meet, fall in love and marry one another over the Internet. The early portions of these electronic exchanges lack what are considered to be the most primary factors for initial attraction: complete physical appearance, age and in some cases lack of voice interaction. Online relationships that start on matchmaking sites, chat rooms and instant messaging are now so common that many psychotherapists in the U.S. devote their practices solely to dealing with problems caused by cyber-romances.

One thing we are also seeing today is computerized matchmaking ability. Many online dating sites have elaborate algorithms designed to match you to a perfectly compatible partner. Some can even have you evaluate photographs to determine the perfect face and body type you desire. These procedures deserve more testing and evaluation to determine their effectiveness.

We can love virtual pets too. The Tamagotchi is a small egg-shaped electronic pet with an LCD screen that fits easily into the palm of a hand. Manufactured in Japan it sold quite well when first introduced 1997 and later in 2005 when a new version came out. Owners press buttons to simulate the giving of food. Users can also play games with the simulated creature. The Tamagotchi beeps when it "wants" something and if neglected can get "sick" and "die", often to the distress of its owner. This toy is able to satisfy basic love needs solely through nurturance, as there are no real aspects of character or personality.

People very easily become attached to such things. Other electronic entities that we can have feelings for include Kismet, a robotic creature with emotionally programmed drives and online chat bots that engage us in conversation. In one study, it was found that people were more self-revealing to a computer program that had itself first "self-disclosed" secret information about itself.

In 2004 the company Artificial Life created a virtual girlfriend called "Vivienne", an attractive brunette that men could download onto their cell phones and then send virtual flowers and chocolates. In return Vivienne disclosed personal information about herself. Perhaps we like such virtual people because we know they are nonjudgmental. They won't criticize us or be mean. Also, we can act in particular ways with them knowing they won't get upset. All of these instances suggest that we will potentially be able to have emotional relations with robots.

What sorts of characteristics might we endow a loving robotic partner? They would need to look human, feel human, be able to think in some ways like we do and be able to express emotions as well as interpret ours. Bill Yeager suggests it would need to have empathy and the ability to converse. To identify with it, our partner would need to suffer some of the same human frailties we do: unpredictableness, perhaps even the capacity to get sick and die. In other words, we would want it to be human. The emerging field that may be able to answer such questions is called robotic psychology. Researchers in this field, called robopsychologists, are devoted to understanding the ways we interact with robots.

CHAPTER 7. A HIGHER CALLING

HIGHER ORDER VALUES

A value is something we act to gain or keep. A virtue as discussed in the social systems section is the means by which we attain values. A value is a goal whereas a virtue is an action or behavior by which a goal is achieved. If someone values justice for example, they can practice it by rewarding their child for taking out the garbage. Then there are motivations. A motivation is the force that drives us to obtain a value. In the above example it would be the desire to do good. The motivation is not the action itself; it is what drives the action.

All animals including humans have built in value systems. We are all born wanting certain things. Across many animals the four basic motivators are thirst, hunger, sex and sleep. These are regulated by feedback loops. Generally, the longer we go without satisfying a basic motivator, the more strongly it drives our behavior. Skipping lunch makes us hungrier which creates an intense drive to find food. Once that food is obtained the drive is sated but then builds up again over time.

Those of us who live in a democratic modern nation are in an environment where we rarely feel thirsty, hungry, horny or tired because these needs can be satisfied in a relatively easy manner. In this case what would next motivate our behavior? The psychologist Abraham Maslow has formulated a hierarchy of needs theory. In his view, whenever we satisfy a set of lower needs (i.e., values) we are next motivated to attain a higher set of needs. Lower more primitive needs occupy the base of his hierarchy, which is shaped like a pyramid. Higher needs occupy levels farther up. A healthy individual could make it all the way to the top assuming each level beneath was satisfied.

In Maslow's pyramid the lowest needs are food and water followed by safety needs for shelter and protection. Then come belongingness needs such as love. Those are followed by esteem needs, the desire to achieve mastery and to be appreciated by others. After this are cognitive needs. These include a desire for education and to understand the world around us. Then there are aesthetic needs, which are a desire to achieve harmony and order. At the very top on the last level is self-actualization, which means achieving those goals that are uniquely yours.

Maslow's theory has been criticized on a number of points. First, the ordering of the needs across individuals and cultures appears to vary. Second, there is no convincing evidence that unmet needs become more important once met needs are satisfied. His theory is nice though in that it forces us to think about what values are and how they should be ordered.

In a future society higher-level values will assume more importance because modernization and science will take care of the lower ones. It is safe to assume that physiological and safety needs will be satisfied in a modern global society. How then may we satisfy the others? Emotionally-fulfilling relationships with colleagues, friends, and family can sate belongingness needs. Establishing and maintaining social relationships ought to be easier in the future as we will have algorithms for matching people based on personality compatibility. This should produce longer and more satisfying social relations reducing social conflict, lowering the divorce rate, and producing healthier children raised in a happy household.

Feeling good about one's self based on accomplishments will satisfy esteem needs. This need is really about excellence. This is important because doing a job poorly produces economic costs. With increased automation, we should see robots and machines take over much of what we do now and do it better than we can. This may result in lowered esteem and a blow to our pride. What good are we if we can't do anything as well as our own creations? Instead of feeling bad about ourselves, this should be an opportunity for us to strive harder at doing those things the machines can't. Also, with enhancements we can improve our abilities to match or exceed those of the machines. With the merging of biology and technology, this ultimately won't be a case of us vs. them, as we will be them and they will be us.

Aesthetic values are the next level in Maslow's hierarchy. These are satisfied primarily through the production and appreciation of art. How will art change in our future? Most people think of art as the last field that computers will excel at but this is not true. There are a number of computer programs now that can compose poetry and music, create paintings and even write novels. Creativity can be analyzed and formalized just as any other skill. Computers may be so good at doing this that they will customize art to our own individual preferences. For instance you could enter a list of songs you like and have a music AI program generate an endless list of songs all of which you will love. The same could be done with movies and literature.

This sort of customization will be one of the major features of our future. Once aesthetic preferences and compatibility are understood we can be matched with anything from the perfect husband to the perfect sandwich. Algorithms may even be able to read our particular moods and desires of the moment and adjust the matching process to suite it as these things do fluctuate within an individual over time. Computers will also allow us to express our creativity in ways we never thought possible. There will be programs that allow us to create realistic 3-D movie or VR simulations, as well as tutorial programs that could help train us on a particular skill. Imagine a program that could watch us paint a picture, providing us with expert feedback as we go along.

Self-actualization according to Maslow is the highest human value, being the ultimate expression of who we are as an individual. Each of us has unique traits and skills. Exercising these is part of what makes us happy. Computer programs will be able to assess what these are and make recommendations about goals we could pursue that would be self-actualizing. For one person this may be teaching. For another it could be competing in a marathon. Self-actualization seems to involve all aspects of the self. It is physical, emotional, cognitive and spiritual. Acting in a way that satisfies all of these is the ultimate happiness.

Values are very important to the flourishing of civilization. Maslow's hierarchy is really just the starting point and should not be taken as absolute truth. There are many other values he does not address, like wisdom and productivity. What about tolerance and diversity? As a society we need to debate values to determine their importance. What others are missing here? Which ones are more important? Should we program values into people using genetic engineering? One can only imagine what humanity could accomplish if we were motivated to excel at everything we do. It could be possible to program a feeling like hunger into a person that would get progressively stronger the longer they went without succeeding, driving them to work harder and think more deeply about what they are doing. Of course this must be a choice and cannot be imposed upon anyone without their consent.

RELIGION AND SPIRITUALITY

One persistent feature of humanity is the presence of religion. It has existed in every known culture since recorded human history. Why is religion so important? A number of reasons have been given. One is that it sets out a moral code prescribing how we should act in society. Another is that is comforting because it assures us that we will live on after we die. Yet another is that explains how the universe was formed and how it operates.

Whatever the reason, religion seems to fulfill some basic need we have as humans. Just as we have basic drives like thirst, hunger and sex, we seem to have a basic need to satisfy spiritual desires. This desire seems the most abstract. It goes beyond basic physiological drives like those just mentioned, or emotional drives like love or even cognitive drives like curiosity. Spirituality seems to be a blend of emotion and cognition. It makes us feel content and peaceful, sometimes even joyful but at the same time satisfies a need to understand what is around us. There is a sense of belonging to spirituality as well, a sense that we are at home in the universe.

The need for spirituality can be satisfied in non-typical ways. One is identification with nature. This explains why many of us like to hike, camp and spend time outdoors. It is liberating to be out among the forests and mountains, the deserts and oceans. There is a serenity to be gained from spending time in the wild. When outdoors we feel small against that size of the natural world. But that is a good feeling. It helps us appreciate the scale of the universe and our place in it.

A second way we can satisfy our spirituality is through philosophy. Whereas religion is based on faith, philosophy is based on reason. Despite these differences they both provide a description of the universe and man's role in it. They tell us what reality is like, how it is we can know it and provide a standard of acceptable and unacceptable behavior. Both also have something to say about the supernatural. Religions posit the presence of universes outside our own, a heaven or hell, inhabited by angels and demons. Some philosophies also split the world in two, with separate realms for mind and body, what is called dualism. Cosmologists who propose the existence of the multiverse can also be included here.

For most people religions are easier to accept than philosophies. They don't require as much thought. The emphasis is on believing which is easier than having to reason. We are also taught religious faiths as children in church and at school. Philosophies typically must be acquired later in life when our cognitive faculties are more developed. Religions are a mixed bag. For the most part they offer a good set of morals: don't lie or cheat, don't steal or murder. But they discourage us from questioning and thinking critically. It is on this point that they pose the greatest danger. When taken to extremes as we see in fundamentalism, they not only discourage rationality but also justify torture and murder of those who dissent with a particular scripture. It is this that is the cause of the current terrorist crisis and historically has been behind inquisitions and wars. As a whole, humanity would be better off with an objective, rational philosophy.

Before concluding it is worth a brief discussion of God. All religions have some version of a God. In most monotheistic religions God is the same. God created the universe, is aware of and understands all that transpires within it and can control everything that takes place. An especially important role for God is as judge. Gods reward the faithful and punish sinners. In other words Gods are omniscient, omnipotent and mete out justice. This version of a God is the ultimate form of anthropomorphism in which we take the image of ourselves and project it out onto the universe. Part of this is a need to explain causality simply. It is easier to attribute natural forces like the blowing of the wind and the movement of the planets to an invisible, human-like hand than it is to understand them in terms of physics and natural laws.

TRANSCENDENCE

Transcendence is a step up to an advanced form of awareness and power. There is a strong similarity to religion here in which the soul at death passes on to heaven where it can live on eternally in the form of a spirit or God. In most faiths these entities have powers that allow them to control what happens in the mortal world. In AI circles it amounts to uploading our consciousness into a computer system where it can live on forever interacting with other minds and possessing a vast awareness of all known knowledge. In biology, transcendence seems to correspond to the creation of an emergent level of organization like the formation of molecules into cells or the formation of cells into organs that can then give rise to things like consciousness.

If neural prosethetics were connected to the Internet we could rapidly access and act upon vast sums of information. If we allowed supercomputers to facilitate this process the result would be an exponential leap in our own intelligence. In effect, we could "outsource" our intelligence using the computers to perform literature searches, analyze data or perform any other computationally intensive task we want done.

Another means of attaining transcendence could involve biological augmentation. Imagine grafting an additional cerebral cortex onto your brain. This could effectively double our thinking capability or it could provide us with the ability to think in some as yet unrealized way. This method could be supplemented with pharmaceutical enhancement, the taking of drugs that could boost our attention, memory and problem solving skills. Transcendence implies more than just an amplification of existing skills though. It would involve a qualitative and drastic alteration in the way we think, one we might not be able to imagine or even understand before it happens.

It is one thing for a machine to be intelligent. It is something else altogether for it to achieve consciousness, that subjective state of awareness many animals and we humans seem to have. If we could build a machine that is conscious it might give us insights into how to boost our own consciousness. Alternatively we may be able to interface and share the machine's conscious state. This shared human/machine consciousness could be another form of transcendence.

In recent years a number of investigators have designed computer programs that they hope will exhibit signs of consciousness. Some researchers have adopted a biological approach, basing their models on the way brains operate. Igor Aleksander's MAGNUS model reproduces the kind of neural activity that occurs during visual perception. It mimics the "carving up" of visual input into different streams that process separate aspects of an object like color and shape. Rodney Cotterill also adopts a neural approach. He has created a virtual "child" intended to learn based on perception-action couplings within a simulated environment.

Other researchers utilize the cognitive or information processing perspective. Pentti Haikonen has developed a cognitive model based on cross-modal associative activity. Conscious experience in this model is the result of associative processing where a given percept activates a wide variety of related processes. Stan Franklin has built a conscious machine model based on global workspace theory. His program implements a series of mini-agents that compete for an attentional "spotlight". The global workspace model uses a theater metaphor of mind in which items that get "on stage" reach conscious awareness.

The question of whether or not humans can construct a conscious artifact remains to be seen. If consciousness or transcendence requires a certain level of complexity, then we may need to wait until humankind develops the technology capable of such complexity. Even if humankind lacked this capability, it does not rule out the possibility that other intelligences do. A conscious artifact thus remains possible in principle. Transcendence may require a form of computational complexity beyond what we see in our brains and our computers now.

COSMIC INTELLIGENCE

Our universe is governed by a set of physical laws that have allowed intelligent life to exist and be aware of the universe. This is a truism since we are alive and intelligent and capable of perceiving existence. But one could imagine alternate realities where this is not the case. Imagine a universe where the laws of physics are slightly different, perhaps there is no such thing as gravity or the nuclear forces that govern how parts of atoms combine is different. If this were the case, the elements we know might never exist. Life, which is dependent on these elements, could thus never exist, so we could not be here to ponder such questions. The universe from this perspective seems "special" somehow, as if it was designed to accommodate us.

This sentiment is known as the anthropic principle. There are two versions. The strong anthropic principle, according to the cosmologists John Barrow and Frank Tippler, states that the universe is compelled to have conscious life within it. The weak anthropic principle is less demanding and according to its originator, the physicist Brandon Carter, states that there is a selection bias for intelligent observers. Only in the sort of universe we have could we make such observations. Implicit in both of these views is the notion that there are multiple universes (the multiverse), each with different parameters, some of which may support intelligent life, others that do not.

Those arguing for the existence of the multiverse have an uphill battle. We cannot and perhaps will not ever know if there are other universes so we cannot be in a position to state what they would be like. Even if there were other universes, we cannot say what their parameters are. Perhaps all universes are equally able to foster life and the distribution of their characteristics is not as wide or as random as we may believe. Let us take an imaginary parameter and say that it must be set to 0.5 for us to exist. What would the distribution of this parameter be across all possible universes? Are all its values (0.0 to 1.0) equally likely? Are the values distributed normally with the average being 0.5 and with lower and higher values less likely? The truth is we cannot say.

Another important point here is that life may be able to exist in other forms in universes with different physical laws. We know that carbon-based life forms such as our own require a universe of the sort we are in. But might life exist in other universes with different physical laws and parameters? Why not? We cannot exclude the possibility. Perhaps there are life forms made up of gases but they exist only in universes that allow certain types of gases to form. Perhaps there are life forms with brains made up of crystal but only in universes that allow crystals to assume a certain atomic configuration. The truth is there may be life forms that exist in forms so exotic we cannot even imagine them because our imaginations are necessarily limited by yes, you guessed it... the laws of our universe.

The French theologian Pierre Teilhard de Chardin postulates that the universe is becoming more complex and more conscious and that it will eventually reach a point, called the Omega Point, where these values are maximized. The Omega Point is not just the destination toward which the universe is developing; it is also the cause of that development. This point is independent of the universe and directs the universe toward itself, in other words it is being driven casually by it. The analogy to be made here is that the Omega Point is God or Christ. According to Chardin, the Omega Point is in existence before the universe and is independent of the universe. It is able to act outside of the restraints of space and time. The Omega Point is also personal and not abstract. It is not just a physical law or property but a being ostensibly capable of consciousness, compassion and intelligence. Chardin says that we conscious human beings are destined to eventually merge with this form of consciousness.

Frank Tipler elaborates on the processing aspects of the Omega Point. He sees the universe as contracting (in an information sense, not a physical sense). As it contracts the distances between points get smaller and communication between those points gets faster. The processing capability thus increases as it continues to shrink. Once it shrinks to a single point the processing capability becomes infinite and the system is capable of simulating all possible futures, what he calls the "Aleph" state. People and other conscious beings can be run as simulations in the Aleph state. Humans at this stage will be omnipotent and omniscient, being able to see all of the past and predict all of the future. One can liken this state to heaven or nirvana.

By now you may have noticed a striking similarity between the Omega Point and the singularity. In both cases we have minds that are either "uploaded" or evolved to the stage where they merge with some greater intelligence. In the religious view this greater intelligence is God. In the AI view it could be a computer superintelligence. In both cases once we have reached this stage we are able to exist virtually forever and able to participate in any type of simulation. The similarities here are not accidental. I think they are both a product of human nature. We humans want to become immortal. We want to be all-powerful. We want to be free to think and experience everything that is. The traditional form of these desires is with heaven and God. The modern take on it is with computers and mind uploading. The theological view with the Omega Point and Aleph State are an intermediary perspective that takes on some characteristics of both. No matter which of these you happen to believe they reveal more about our own nature than they do about the future or the universe.

CHAPTER 8. SOCIAL SYSTEMS

VIOLENCE AND WAR

A history of our species reveals one all to obvious feature: war. It seems to be one thing that we are really good at. We also never seem to learn from our past mistakes and have been unable to rid ourselves completely of this scourge, no matter how wide the death toll or horrific the atrocities. An evolutionary account of war explains it in terms of competition for scarce resources. Killing off the competition means more land, food and other things for one's own group. Whereas war is a social phenomenon, its seed is an individual one. It stems from certain aspects of our psyche that include aggression and the lust for power.

Violence and aggression are part of human nature as we know it now. Should genetic engineering become possible we may be able to eliminate or at least reduce some or even all forms of aggressive behavior. Few people would object to the end of bullying, theft, rape, torture or murder. But we should be careful not to throw out the baby with the bathwater. There may be aspects of aggression, such as confidence, ambition and competitiveness that we want to keep. A peaceful society would be a utopia but not if we were all made bland and sheepish. Connected to aggression may be the need to excel. This may motivate us to overcome obstacles and succeed. If this drive could be preserved without the desire to cause harm to others it could accelerate the rate of human accomplishment.

A future in which humanity is less violent would mean that gun control is no longer an issue. People in this scenario may only use guns and other weapons when dealing with rogue individuals, wild animals or other potentially dangerous beings. Since laws are mainly in place to deal with people who act harmfully, the end of violence may see the elimination of many laws restricting people's behavior. Police and courts could vanish or at least be minimized, so there would be a reduction in the size of government. A future political order without violence fits a more libertarian vision of governance.

This may seem strange but in some sense we humans have survived and flourished in part *because* of war. War at its heart is a form of large-scale social cooperation. It calls on many members of society to perform specialized actions in a coordinated way to achieve the difficult goal of defeating one's enemies. The result is that one group dominates and gains access to limited resources, ensuring their survival. War also motivates the development of new technology to build more effective weapons. If we could take the engines of war and harness them toward good instead of evil, we could construct new dams, bridges and cities rather than weapons. We could do this through large-scale social cooperation directed to end global problems like hunger, disease and energy needs. How this gets implemented is complex. We need to be weary of governments orchestrating these activities for us and rely on free market mechanisms.

It is all too easy for us to be seduced by power, defined here as the need to dominate and control others. Paired with aggression the quest for power has probably caused more harm to human society than any other psychological trait. At a small scale it contributes to spousal and child abuse. At a larger scale it produces military coups, dictatorship, oppression, war and genocide. The unequal distribution of power in society contributes to and is a result of negative stereotyping, racism, sexism and classism.

How can we put a stop to all of this? One way is to spread or distribute power more equally. Democracy does this through self-governance. The balance of power between an executive, congressional and judicial branch of government is a system of checks and balances. This helps but reforms are needed, as in the U.S. history shows that presidents rarely seek congressional approval before the deployment of troops abroad. Capitalism as an economic system also helps as it reduces the influence of government in the economic sphere.

However, I would suggest another less considered alternative and that is to model aspects of human society on systems that already achieve a balance of power, namely natural ecosystems such as we see in coral reefs or tropical rain forests. Patterns of predation, energy flow and symbiosis in ecologies can serve as examples for how we may balance power amongst ourselves. Species competition for instance can be taken as a proxy for competing corporations or social classes. Discovering how "mother nature" has solved such problems may provide key insights into our own social issues.

Another way to lessen the risk of war is through education. We need to be more truthful in the telling of history than we currently are. Many nations gloss over the less inspiring aspects of their own history. For example, the Japanese have avoided a full accounting of forced prostitution during World War II, the so called "comfort women", or of the atrocities committed in Nanking in 1937. How many of us know that the Ottoman Turks murdered as many as 1.5 million Armenians and other ethnic groups after World War I, many through death marches into the Syrian desert? How many know that a half million Tutsis were killed, many by machete, by the majority Hutus in Rwanda in 1994? It may not be pleasant but we cannot let ourselves forget such incidents. The young, at an appropriate age especially need to know of the barbarity of human nature so that future generations do not make the same mistakes.

There are steps in the right direction and there is cause for some guarded optimism. The International Criminal Court was formed partly in response to the Rwandan genocide. Global statistics that measure violence have also been on the downturn for many years. There are fewer people now who die in interstate and civil wars, homicides and hate crimes. In addition there has been an increase in civil rights legislation for groups such as children and homosexuals. Stephen Pinker has compiled and described many of these statistics in his book *The Better Angels of our Nature*. Matt Ridley also makes this case in his book *The Rational Optimist. How Prosperity Evolves*. We are now better fed, clothed, sheltered, protected against disease, and entertained than at any time in our previous history. By balancing power and seeking rational democratic solutions to problems we hope it can continue.

REPUBLICS AND EMPIRES

In the many thousands of years we have existed it is only recently that we have had representative government and stable political transitions. Most of our history has been characterized by tyrannical rule by a power elite, whether that be a king, dictator, general or priest. The citizenry are the ones who have paid the price for this, alternately being massacred, enslaved, or forced to fight in endless wars.

Something amazing happened beginning in the 18th century. The United States and other European nations began to democratize. They created parliaments and congresses. There were elections where the common people could vote for the candidate of their choice. This form of government is a democracy or republic. Of course we are far from having global democratization now. There are still many countries ruled by military juntas and despots, but they are starting to become the exception not the rule. It is expected that the number of democratic nation-states will continue to grow and that this will be the standard form of human society in the future. A recent estimate has about 60% of the world's nations currently being democratic.

Generally speaking democratic nations fare better than others. They have protection from abuse of power due to checks and balances. There is limited majority rule, defense of individual liberty and legal protection because all citizens are guaranteed the same rights. Democracies are far from perfect of course and suffer from corruption and election fraud among other problems. However when compared against their alternatives like socialist dictatorships or rule by inheritance, privilege or force they are clearly better.

An empire by contrast is an assemblage of states and peoples who are united and ruled by a central authority, either an emperor or oligarchic ruling class. Territory and control is acquired by these organizations through force and military conquest. People in empires do not have a say in how they are governed and dissent can be punished by death. Throughout history there have been far more empires than republics. In rough chronological order these would include the Egyptian, Assyrian, Persian, Greek, Roman, and Islamic empires, not to mention the empires of the Far East such as those in China or the European colonial empires or the modern socialist empires of Nazi Germany, Italy and Japan. In fact prior to the modern period true democracies were very few and far between. Certain periods during Greek and Roman civilization have also qualified as being democratic.

An empire is characterized by force imposed from without, a republic by force imposed from within. Empires are ruled from the top-down while republics are bottom-up. Order in empires is typically forced on the citizenry whereas self-organization is more prominent in republics. So in each case we have opposing mechanisms, one case where individual autonomy and rights are trampled and the other in which they are protected. The transition between these two systems is perhaps the most important movement in human history. It is one in which individuals are at last considered more important than the group, where your class, wealth, ethnic background or religion doesn't matter and where everyone is treated equally.

What is the future of democracy? If the trend continues we will hopefully see more nations transition to democratic rule. There are encouraging signs. One of the advantages of globalism is rapid mass communication. Oppressed people, such as those in Iran, can see the benefits of democracy. They can no longer be kept in the dark and lied to by their leaders. They can see on cable TV and the Internet the way the rest of the world lives and desire this lifestyle for themselves. This very same communication makes it easier for groups to organize and hold rallies. During the Arab Spring movement we saw students in Egypt using Twitter, texting and other forms of social media to successfully stage a political revolution. The distributed nature of the Internet makes it difficult to control. Other examples in which technology has helped promote freedom and democracy are the websites Ushahidi, WikiLeaks and World is Witness to report human rights abuses and Enough is Enough to help people vote in elections without the influence of any political party.

An informed electorate makes better choices and information about political candidates and parties is now everywhere. It is difficult to claim you didn't understand your candidate's stance on gun control or foreign affairs when his or her website is just a few clicks away. Beyond this the World Wide Web now has an abundance of different viewpoints. A concerned voter can read editorials from every political perspective, whether it is libertarian, socialist or republican. Many people have blogs and can participate in online chat rooms if they wish to debate specific issues. Information is power. Power to the people!

MONEY AND ELECTRONIC EXCHANGE

Will we need money in the future? The only situation where it could be eliminated is under conditions of unlimited resources. Money after all is a medium of exchange. We use it in order to obtain goods and services that we cannot or will not provide for ourselves. So the question really becomes will we ever achieve a state where there is no want, where we can get everything we desire without having to work for it?

There is one scenario that may satisfy this condition. In the section on nanotechnology we mentioned nanoassemblers as devices that can create nearly any object starting at the molecular level. If these nanoassemblers become easy to make and are in widespread use in society, perhaps there may be a reduced need for money. The science behind such assemblers is still so far into the future and the costs in producing and maintaining such machines seems high. They may also be limited in the complexity or size of the objects they can produce. In addition the raw materials they require may limited, especially under conditions of high demand.

Assuming we continue to use money in the future, in what ways might it be different? One change already being implemented are e-money transactions using mobile devices. The Apple Pay system involves holding the iPhone near a contactless reader with a finger on a Touch ID. The ID is a security measure that reads the user's fingerprint to verify their identity. Money from a linked credit or debit card is transferred to make the purchase. Receipts can then be generated electronically.

This process will supposedly speed the check out time at a register since one does not need to take out a wallet or fumble with bills, coins and receipts. Apple claims that these e-transactions cannot be hacked because each device is assigned a unique number that is encrypted and stored on a chip on the phone itself and not on a server. The company additionally claims they will never share purchase history or security information with other vendors or merchants.

If this payment system comes to pass it will mark the start of a new stage in the use of money. Before money there was a barter system in which items deemed to be of equivalent value by traders were simply exchanged: one bracelet for one sack of corn for example. Then money both in the form of coins and then as bills was used. This was followed by the first incarnation of e-money in which funds could be wired between banks and vendors through the use of credit and debit cards. Then the second incarnation of the e-money system by mobile device is the stage that is appearing now. What will follow this? Perhaps electronic exchanges operated by neural interfaces?

The introduction of the Bitcoin system is another recent change in how we may use money. Bitcoin is a currency created in 2009 in which transactions are made solely between the seller and the buyer. There is no middleman or bank involved. There are also no transaction fees and no requirement for the buyer to provide his or her real name. International payments are supposed to be easier under this system because the Bitcoins are not tied to any country and are thus not subject to regulation. They can be bought on an exchange using different currencies and can be transferred between individuals using mobile apps or computers.

Bitcoins are stored in a "digital wallet" that is either on the user's computer or in an account in the cloud. This serves as a private "bank" to which Bitcoins can be transmitted or received, used to purchase things or simply saved over time. The FDIC does not ensure these accounts and should there be a run on the market, investors could potentially stand to loose their money. Regrettably, Bitcoin cloud wallet accounts have been hacked and companies have stolen client's money. Because the coins exist solely in a computer they can be deleted accidently or destroyed by viruses. The anonymity of the system has appealed to those who sell drugs and engage in other illegal activities.

It is not clear whether the Bitcoin system will continue in its current state or be regulated by governments seeking to collect tax revenue and to control the currency's value. There is no central bank or Treasury that creates Bitcoins. Historically, when the U.S. government has needed to pay for something, it printed more currency. The increase in money put into circulation this way leads to inflation and currency devaluation. So the success of this system depends upon your perspective. Those favoring lower taxes and less inflation will argue that Bitcoins increase economic growth and lower unemployment. Many of these same individuals will argue that the so-called illicit activities (drugs, prostitution, guns, etc.) proliferating under the conditions of anonymity ought to be legalized anyway and so additionally contributing to economic growth.

THE FUTURE OF CAPITALISM

Michio Kaku in his book *The Physics of the Future* discusses how wealth may change in the coming century. He starts by describing three bubbles and crashes experienced by the economy over the last 200 years. Each of these bubbles began with a technological invention that was followed by innovation and progress. Then there was overspending and speculation that led to an economic crash. The first bubble was steam power used to create engines for locomotives and boats. This fed the industrial revolution that increased societal wealth tremendously. But locomotive companies were overinvested in and by 1850 there was a crash that slowed the economy.

The second bubble came in the early 20th century and was led by electricity and automobiles. Autos were now overinvested and for this and other factors as well, there was another crash, this time the Great Depression of 1929. The third wave of technological innovation was in the form of computers, the Internet and electronics and began in the 1980s and 1990s. The corresponding crash happened in 2008 and has been dubbed the Great Recession.

What will the next wave be? Kaku speculates that it will come in the form of artificial intelligence, nanotechnology and biotechnology. There is a lesson to be heeded here. Investors should be wary of putting too many of their eggs in one basket. Optimal investing involves a balance across many sectors of the economy, across large and small companies, across domestic and foreign capital and across monetary instruments like stocks and bonds. Overinvestment in one area like tech companies is a recipe for disaster. We also need to be sensitive to the timing of these bubbles and crashes to anticipate when the next one may occur.

Kaku in addition lists four major ways technology will affect the future economy. The first of these is perfect capitalism, a condition that exists when the producer and consumer have infinite knowledge of the market. The information age is now allowing this to happen. Customers can go online and research products and do price comparisons. Producers can track shoppers buying habits. Together these help to make the perfect price at which supply meets demand. The second major effect is customization by which producers can tailor a product perfectly for each individual consumer, such as a shirt that fits perfectly. It is currently too expensive to do this in any mass production way now, but increased information and automation may make it possible down the road.

Technology as a utility is the third effect. In this scenario computation and tech services will become so cheap and ubiquitous that they can be treated as utilities, like water and electricity. We will access computation and data off the "cloud" as we need it and only pay for what we use. There has certainly been a trend over recent years to see a migration of software and data off of our hard drives and into the cloud. Customer targeting is the fourth effect. In this view, service providers will have increased information on customer's buying habits and preferences so that they can target specific products to specific individuals.

Imagine walking by a storefront window on your way to work. A sensor in the store scans your gait and facial pattern and identifies you. It then projects an audiovisual hologram into the air in front of you showing the pair of designer sunglasses you have looked up several times online. It says that if you stop in now, you can buy them for 20% off. You look at your watch. You are little early. Oh what the hell, you duck quickly into the shop to buy them. This scenario is probably less than a decade or so away.

Robots will also play a role in the coming economy. Robots have traditionally been confined to the three D's: work that is dirty, dull and dangerous, although this is changing and robots are now capable of assuming a greater variety of roles. Advances in robotics will likely see them move from the factories into our homes and from performing industrial work to doing household work. Robots could be put to a myriad of uses doing the sorts of things we don't like to do. Tasks like vacuuming the rug, doing laundry and taking out the garbage are likely candidates.

A major concern over technology relates to its effect on the economy. The fear is that robots or other automating devices will replace human workers, contributing to unemployment. It is true that assembly line robots and robotic arms have supplanted workers in automotive factories and may continue to do so. This can result in temporary human labor displacement. But cost savings at these companies should produce increased profits that can lead to a greater number of jobs in the long run. If a company grows prosperous by utilizing robotic workers then new plants can be opened, employing more workers. Also, there is a whole new industry needed to design and maintain these robots that will spring up employing more workers.

The Internet has additionally had a transformative effect on the economy. It has helped to eliminate the "middle man" by making information and action directly available to the consumer. Twenty years ago if somebody wanted to go on vacation they would need to see a travel agent who would research flights, hotels, and sights. Now these same functions can be performed in minutes on the World Wide Web. We now shop and purchase nearly everything online, saving us time from having to drive the car to the shopping mall. This has probably reduced the number of "brick and mortar" retail stores but also expanded these same businesses with an online presence.

ENERGY AND THE ENVIRONMENT

One of the most heated debates in contemporary society is on energy use and how it affects the environment. Older energy sources like coal and oil produce carbon dioxide that builds up in the atmosphere trapping in heat, the so-called "greenhouse effect" that contributes to global warming and climate change. Scientists generally agree that the world's temperature's are on the rise. In addition to the increase in temperatures changes can be seen in the shrinking of the polar ice caps and sea level rise. This could have a devastating effect in the coming decades, flooding coastal areas, potentially killing many people and animal species, affecting food production and the global economy.

Scientists also mostly agree that the increase in temperatures is tied to carbon dioxide. Ice core samples show they are correlated, rising or falling in synchrony over a several thousand-year period. Within the last century there has been a sharp spike in both of these measures, which would correspond to the increase in man-made carbon dioxide output. Natural changes in these levels that came prior to the industrial age tend to be slower and more gradual. Computer simulations and measures of heat radiated from the Earth's surface also support the human activity-induced hypothesis. Our goal for the future then is to reduce human carbon dioxide emissions worldwide. This will hopefully bring temperatures back down and prevent some of these catastrophic effects from taking place.

There are some who argue for energy reduction and the more efficient use of existing resources, hence the phrase "reducing our carbon footprint". A number of corporations have adopted this stance. For example in 2007 Walmart reduced its energy consumption 20%-30% by painting the roofs of their stores white to reflect more heat, adding skylights to substitute for artificial lighting and enclosing its refrigeration containers. Other demands to "go green" include the use of compact fluorescent light bulbs that have yet to go into wide usage.

Although such measures can be helpful they ought to be used as a stopgap measure only until other alternative energy sources become available. There is more than enough energy on this planet to supply all of our energy needs and we have the technology available now to utilize it. Once we are able to tap into this energy, we won't need to cut back on consumption. Progress requires increased use of energy and we should focus on the demand side of this equation, not the supply side. The best alternate energy sources are clean, cheap and renewable. There are a number of environmentally friendly options available. These are solar, wind, hydro, geothermal and nuclear, in which power is obtained directly from the sun, wind, water, earth, or radioactive sources.

Currently the price of solar panels is expensive and so solar energy has not been used on a large scale in many countries. Families who can afford them install such panels on their roof and in the long term will see cost savings. New technology has been bringing the price of these down and by some estimates they will be as affordable as coal and oil within the next decade or two. Germany uses the most solar energy, followed by China, Italy, Japan and then the United States. A potential future for solar power is to launch space satellites into orbit that absorb the sun's radiation and then beam it back down to Earth using microwaves.

Power from the wind is another clean energy source. Wind farms made up of turbines are in use around the world. They can be located either on land or offshore. A single wind generator can produce enough electricity to supply a small village. One hundred windmills create the same amount of energy produced by a coal or nuclear power plant, about 500 megawatts. They do need to be located in areas where winds blow at a fairly constant rate and need to be close to where the electricity will be used due to transmission loss along the wires. Some people consider them an eyesore.

Hydroelectric power is generated by falling or flowing water and is now the most popular type of renewable energy. It makes up about 16% of global electricity creation. It is usually harnessed through the construction of dams such as the Three Gorges Dam in China, the largest in the world, generating 22, 500 megawatts. Once construction costs are accounted for it is very affordable and produces very little carbon dioxide. Downsides include disruption to local ecosystems and wildlife.

We are occasionally reminded of the immense energy beneath our feet when a volcano erupts. Much of the interior of our planet is molten and creates an incredible amount of heat that can be tapped into and used. Unfortunately, electric generation by this method is relatively rare as it has been limited to locations near tectonic plate boundaries. As of 2012, only 24 countries were using geothermal energy. Compare this to the 150 or so nations that are using hydroelectric power. Geothermal power has the greatest potential for development but costs for exploration and drilling are expensive.

There are two types of nuclear power. The first, called nuclear fission, creates energy by splitting uranium atoms. The major cost of this is that large amounts of nuclear waste materials are generated as a byproduct. These pose a hazard for thousands to millions of years and as of yet, there has been no good place to store them. A core melt down is also a possibility. This is what happened at the nuclear power plant in Chernobyl, Russia in 1986 that spewed radioactive particles over large areas of the former USSR and Europe. These problems make it a fairly unpopular choice as a future energy. In the U.S. the last reactor plant was constructed in 1977.

Michio Kaku outlines several exotic energy generation sources that could be put to use in the next century. Among these is the second type of nuclear power called nuclear fusion. In this process hydrogen atoms are fused using heat and compression. This generates a large amount of energy with very little waste. Methods being investigated now involve focusing many high-powered lasers on a hydrogen pellet and using magnetic fields to pressurize hydrogen gas.

One final energy source worth mentioning is magnetism. Room temperature superconductors can be used to elevate cars or trains that can then move forward with very little friction. They can be propelled to high speeds with only an initial energy "push". The MLX01 train in Japan in 2003 set the world's speed record at 361 miles per hour. Within a single generation we may see personal cars floating on superconducting magnetic transportation networks.

CHAPTER 9. CIVILIZATION

STAGES OF DEVELOPMENT

A number of theorists believe humanity has passed through several important stages in its history. The stage at one point in development serves as the antecedent for the next, setting up its necessary preconditions. People can agree more or less on what these stages might have been. The real speculation starts when we project these forward into our future. The events in these stages, at least those for which there is historical data, show an exponential increase, meaning that the time separating them gets shorter and shorter as history progresses. In this essay I will describe some of these proposed stages and comment on them.

The standard view on humanity's progress is that there have been three major stages and that we are currently in the fourth. The first was the hunter-gatherer stage during which we survived by a supposed division of labor in which men went out hunting to catch game and bring it back to the tribe. Women would have stayed at home and cared for children while going on shorter forays to gather things like nuts, roots and berries. This separation could have led to differing skills with men evolving superior navigation skills and spatial reasoning. The women would have evolved better linguistic and verbal skills as they would have spent more time vocalizing and communicating among each other and with children. Such cognitive differences do exist but they are small and only become statistically significant when large samples are tested.

Stage two is the development of agriculture. Because groups now had to plant seeds and care for crops, they had to stay in one place. This may have been the catalyst for the emergence of cities and the further development of social cognitive skills since we were around larger groups of people for longer periods of time. It also saw the development of more complex social classes like political leaders who governed, generals who waged war and religious figures who preached and established a moral code.

Following this was the third stage corresponding to industrialization. Here machines were used to produce products that were normally manufactured by hand. For the first time we built things that could build things for us. The dominant cognitive skill that may have evolved here was the notion of the world as mechanism. In this view all objects natural and artificial can be explained by breaking them down into a set of interacting parts. This understanding fostered more complex feats of engineering and architecture as well as the modern conception of science based on causality.

The final stage was the information revolution that has occurred essentially within the last century. The generation and dissemination of information characterize this time period. It has started off slowly but has been accelerating at an exponential pace. The cognitive and social changes accompanying this period involve the use of information to accomplish every task from communicating with others to shopping.

Ray Kurzweil outlines another stage theory of humanity, only his stages, which he calls epochs, are all concerned with information. The first epoch is characterized by the representation of information in physics and chemistry followed in the second by information concentrated in DNA. After this information is represented in neural patterns, essentially the thoughts inside brains. After this in epoch four, information is situated in computer hardware and software. This brings us through to the present day. The next two epochs are predictions. Epoch five involves the merger of technology and human intelligence, essentially the integration of brains with computers, where the differences between biology and technology disappear.

The final epoch, number six, is the most controversial as it involves the saturation of matter and energy with information. At this stage the universe is said to "wake up", what might in a religious interpretation be the birth of God, or at least of God's mind. There is a related concept here, which is that of the Omega Point. Developed by the French Jesuit priest Pierre Teilhard de

Chardin, this is a point of maximal complexity and consciousness toward which the universe is said to be evolving.

A Russian astronomer named Nikolai Kardashev formulated the final stage theory we will examine here. This scale measures a civilization's level of technological development by the amount of energy they use. A type I civilization is capable of harnessing all the energy of its home planet. Possible energy sources here include fusion, antimatter and renewable energy. Type II civilizations can utilize all the energy of their local star. A proposed method of doing this is by constructing a Dyson sphere or one of its variants. We discuss Dyson Spheres in the essay on megaengineering. A type III civilization can make use of all the energy in its galaxy. One method by which this might be accomplished is by tapping into the energy of a galaxy's supermassive black hole. These are titanic feats of engineering and the amount of time spent separating them could be in the thousands if not the millions of years. One estimate is that humanity will reach type I status in 100-200 years, type II in a few thousand and type III in a million. These are probably conservative estimates.

Other benchmarks to measure humanity's achievements include the amount of available information on a letter scale which can range from the letter A representing 10^6 unique bits of information to the letter Z, representing 10^{31} bits. This comes from Carl Sagan, who puts humanity currently at a 0.7 H. John D. Barrow has an alternate classification based on our ability to manipulate matter at increasingly smaller scales, what he calls Type minus stages. These range from Type 1-minus to Type VI-minus and beyond. These levels span the range of basic manipulation of solids to the manipulation of the smallest particles of matter. In this scale we are estimated to be between a type III- and a type IV-minus level. Robert Zubrin classifies civilizations based on how widespread or dispersed across space they become starting with planetary, then followed by stellar and galactic dispersion.

Most of the stages so far described use quantitative metrics like information, energy or space to assess development. It is easier to make predictions based this way because one can scale upwards by known amounts. What is less easier to anticipate are qualitative changes. For instance there may be some newly discovered way to modify the human brain that produces a fundamental change in the way thought occurs. If that were the case it would be a "game changer" and it would be difficult to say what type of large-scale changes would ensue from that point on. Likewise, the discovery of a new type of matter or fundamental particle might completely reorganize the way we understand physics leading to radical alterations to society that are not easily predictable.

The way in which humanity might react during the transition from one stage to another is also difficult to say. At a very broad level we may anticipate those who oppose the adoption of new technologies or discoveries, as was the case with the Luddites at the start of the industrial revolution. These people will point out the downsides of the new technologies that may include damage to the environment, jeopardies to human health and alterations to the nature of what it means to be human. At the other end of the spectrum there will be people like the transhumanists who welcome the advancements. Moderates will likely oppose certain changes but agree with others based on their own political and social viewpoints.

PROGRESS AND CREATIVITY

Will the future see a continuation of the tremendous progress we have seen in the last century? Will there continue to be the rapid improvements we see taking place in science, engineering, technology and other areas? What factors would promote this continuation? Virginia Postrel in her book *The Future and its Enemies* argues for the existence of two groups with opposing views. The "statists" are those intellectuals and politicians who are opposed to change. They argue that technology limits the human condition, that economic change produces instability, that popular culture is unrefined and that consumerism pollutes the environment. The statists would like to see a managed future in which the state regulates these endeavors, often according to their own particular agendas.

In contrast are those in support of what she calls "dynamism". These people favor an open-ended society where creativity, free markets, entrepreneurship and innovation are allowed. She provides examples from many different areas where a natural trial and error process gave rise to new and unpredictable creations that benefitted society. In her view freedom with rule of law are necessary preconditions for the advancement of the human condition. There is no one enforced way of doing things in these societies, they are characterized by competing interests and so are pluralistic.

Earlier scholars have made similar arguments. The Austrian Economist Joseph Schumpeter in the early 20th century saw history as a series of upheavals in which bouts of "creative destruction" regularly occurred. In his perspective a new innovation would often produce temporary unemployment and disarray but this would be followed by a period of stability and increased economic growth. As an example horseshoe and carriage makers temporarily lost their jobs after the start of the automotive industry but then were rehired and began to make more money as the economy reorganized.

Risk is the operative principle here. We need to be willing to take risks in order to see benefits. As the saying goes: "You need to break an egg to make an omelet." If we try to create a society that is without any risk the result will be stagnation. There is a similar principle in physics. Dynamical systems that generate variety are "far from equilibrium", they have constant energy inputs and are inherently unstable but exhibit variety and novelty. Most living creatures and ecosystems are in far from equilibrium conditions. When an animal dies it returns to a state of equilibrium. By analogy a society in equilibrium is dead. We need to accept a bit of chaos and risk in exchange for progress.

Creativity and productivity drive society forward. We need people who can see things differently and then work to make them come true. Without new ideas and the creations based on them there can be no advancement. Technological developments along with global communication exchange make it far easier now to express and share our creative voices. Here is a case study from the arts. A budding musician uses music software and synthesizers to compose songs rather than having to buy expensive instruments. She then distributes these songs by uploading them to a file-sharing site where they can be listened to and purchased. This is in sharp contrast to the way the music industry used to operate, with a few large record companies who selected artists based on popular demand and produced only music they deemed worthy.

Under the old system, music was more homogenous in style and there were far fewer choices. In fact the music industry was characterized by a single style that dominated for a decade or more – examples include rock, alternative, grunge and hip-hop. Now we see lots of smaller bands producing a greater variety of styles. Rather than record label executives making the decision about what gets produced, we let the consumer have a say. We can independently post reviews online or tick off the number of stars we think a work deserves. This feedback, known as crowd sourcing is a much more democratic way of doing things.

With further advancements in software and information sharing we can expect greater individual creative expression. Imagine a world in which each of us can make our own video games, create our own feature length movies or write our own novels. We could then market and make these available online. It is also worth mentioning that these same advances make it easier on the consumer side to search and locate the type of art we like. Search engines are a start but companies like Netflix, Pandora and Amazon have all created algorithms to easily recommend works you may like based on previous buying choices.

HAPPINESS AND CIVILIZATION

There has been an explosion of social science research on happiness in the last decade. What was formerly within the discipline of philosophy now fits more squarely in the realm of psychology. We know for example what factors correlate with self-reported happiness and which ones don't. We also have roadmaps for how to get happier in ten "easy" steps. My comments here however will be general and concerned not with how to be happier but about the importance of happiness for the individual and the relationship between individual happiness and society.

Happiness is the main purpose of life. It is the ultimate goal toward which we each should strive. It is very elusive however and each time we seem to have it, it somehow fades or slips from our grasp. This is the nature of happiness. It is not meant to last. Like other emotions and psychological states it is short-lived. The trick is to focus on the process of finding happiness and not the outcome. If we engage in acts that make us happy and not on the end state of happiness itself we will be maximizing the amount of time we can be happy.

Another feature of happiness is that it is different for each of us. What makes one person happy might make another person miserable. We each vary in our traits and proclivities and these dictate the course of action we take in life. Of course there are certain fundamental conditions that must be realized before we can pursue happiness. These include things like health, a minimum standard of living, some degree of economic prosperity and so on. Once these are satisfied though, what is left is the road that only an individual can follow. One of the challenges of life is finding this road. Some people may discover it right away. Others may never discover it and spend their entire lives searching. This is not necessarily a bad thing as it means trying out different solutions to see if they work. Variety is the spice of life. One could do worse than to take on different challenges and experiences and learn from them. It makes for a more interesting existence.

A key component to happiness is freedom and opportunity. If one is to figure out what makes them happy they must have room to explore. They must be able to study a variety of different courses in school or try on differing jobs to see what they like. For this reason it is crucial to live in a free society that allows for this sort of choice. If we are forced to take on the family business or directed by the state into a particular profession then one can almost certainly guarantee a miserable life.

A suggestion is in order here. One must both like what they do and be good at it if they are to be happy. If someone liked a profession but didn't perform well they will be laid off. Conversely, if someone disliked a profession, even if they were good at it, they would probably quit sooner than later of their own accord. Ability in some sense seems to be the same as desire because most of us like to do the things we are good at.

What is the relationship between the individual pursuing happiness and the larger society? Well to start, a civilization is exactly this. It is a collection of individuals who are free to search for happiness. The reason governments exist is to allow us to do this. The moniker "Ask not what your government can do for you but what you can do for government" has it exactly backwards. The government serves the individual not by giving us the things that make us happy. Rather, it should provide the background conditions necessary for the individual to achieve these things on their own.

Heterogeneity in a population is important. Just as there are a variety of people who need things, there are a variety of people to supply them with their needs. In the process of natural selection diversity in an animal population is necessary to adapt to changing circumstances. The same goes for human populations. As long as there is enough variety among people there will be a sufficient diversity of interests to ensure that society can function and adapt to new environments.

One could argue that in the future there could be a convergence of interests, with most people volunteering to be more similar to one another. For example, most of us may not want to perform technical or tedious tasks like accounting or database management. We may instead prefer to engage in professions that are more open and creative. There is nothing wrong with this narrowing if the gaps can be plugged by other means. Automation has the promise to do this. In effect we can "outsource" to automation those things we don't want to do and pursue those we like.

We are of course seeing this already. The Internet has for years been pushing out the "middleman" because it can put consumers into direct communication with service providers, hence the decline in travel agencies and bank tellers. Taken to its extreme we could have a society in which everything is performed by machine supposedly leaving us with nothing to do at all. Humanity in this thought experiment will have lost its uniqueness and will be left as an "inferior" species while machines pave the way to the future.

Is this really such a bad thing? First of all most people would not object to having more time to spend on activities they enjoy like being with family or friends. Second and more importantly, it is likely that we will evolve along with our machines, being more closely integrated with them. In this future, we will continue to take on greater challenges as part of a hybrid biological/technological society. For a more in depth discussion of these outcomes see the section on technology.

SEPARATION AND CULTURAL FLOURISHING

Geopolitics argues that geographical factors play a large role in determining the fate of nations. A mountain range or an ocean can be a more effective defense against foreign invasion than a powerful army. Historically nations like Ancient Greece during the Persian Wars, Japan against the Mongolian hordes and Britain versus Nazi Germany in WWII have been spared costly invasions because they are simply difficult to get to. Similarly one can argue that colonial revolutions are more likely to succeed when separated by large distances.

If we take this argument into outer space it seems to hold true. Distances in outer space are much greater than those separating continents and the technical hurdles needed to transport large armed forces across them are enormous. Consider how difficult it is just to send a few men to the moon or to maintain a continually manned space station in Earth orbit. Next consider the difficulties in projecting military force across stellar system distances such as that between the Earth and Mars. It seems in these cases that the advantages lie with the defenders.

This means that human space settlements may have a high chance of success should they decide to become independent of their founding nations. However, this conclusion is based on several presumptions. First, that advanced technologies are not developed that will enable us to bridge these distances and second that the colonies in question are not overtly dependent on their founding members for economic resources. Those settlements that are farthest and the most materially self-reliant seem to stand the greatest chance of developing political sovereignty. There are other mitigating circumstances. A confederation or alliance of several outlying colonies that trade and support each other stands a better chance of resisting an opposing force.

According to the Fermi paradox any intelligent civilization that developed in our galaxy has had ample time to explore and colonize it. The absence of any such presence supports what I call the isolation principle – that our universe seems designed to prevent the spread of life and to keep alien species separated from one another. This may not be such a bad thing as it makes it hard for any one species or members of one species to propagate and dominate others. Isolation in a sense encourages self-reliance and independence.

Cultures "on the edge" are those that would seem to flourish the most. This refers to colonies or nations that are far away but not too far away from a parent civilization. These cultures are close enough that they can engage in trade and other forms of exchange but far enough away to be creative and challenge existing ideas. The Ionian Greeks had this sort of separation from Athens and the American colonies had it from the British. There may even be a mathematical formula that predicts cultural flourishing based on the amount of time it takes to transmit information and physical goods back and forth. The right amount of isolation may be a good thing.

All of this begs an interesting question: When is it "right" for a nation or political group to secede? Are there cases where a civil war is justified and the rebels should be allowed to win? Libertarians allow for more fluidity in national affairs. If a state democratically votes to separate from a larger governing body then it should be allowed to do so. Under similar circumstances a state could be allowed to rejoin politically with a former partner. In other words, there should be both freedom of entry and freedom of exit between nations and states within nations.

We already see this when it comes to free trade agreements like NAFTA where economic terms can be negotiated. The important principle here is sovereignty. A group of people of a given size can voluntarily step into and out of relations with others as they see fit. Groups will naturally want to join profitable flourishing states and leave restrictive or overly regulative ones so that in the end, good policies end up being rewarded. Patterns of emigration and immigration operate on this notion, in which people have been said to "vote with their feet". Similar ideas have been proposed for private, competing currencies.

UTOPIA

Is a utopia possible? Can we create a perfect life for ourselves where we are all happy and able to pursue our own ends? This is the ultimate goal of a society and ostensibly what every benevolent government or political leader would like to see. However this may not be a likely outcome unless we change human nature. The reason we have laws, courts and police is because people act unethically. If nobody stole or murdered we would not need to pass laws forbidding such actions. Were we to implement a genetic engineering program to stop these behaviors then we would see a better society, but not necessarily a perfect one. There are dangers in doing this however, because in removing such traits we may take away some good things too, like the drive to succeed.

Another reason we may not achieve utopia is that people may desire conflict and challenges. If everything is made too easy we may become bored or degenerate. Under such circumstances people could overindulge by stimulating themselves using drugs or artificial world simulations. This might not be as bad as it seems if the technology at that time is developed sufficiently to prevent addiction and harm. If there are a diversity of personalities in such a future society then there will likely be individual differences.

Some people will seek hedonism and stimulation while others will decide to be more creative and productive, pursuing their own means to happiness.

One way to create a near-perfect society is to empirically compare different societies using demographic statistics and implement public policy changes to see if things change for the better. The first part of this has been done for some time now, with sociologists measuring things like educational attainment, income level and health outcomes between countries. Unlike an experiment though we cannot directly manipulate one society while leaving the other as the un-manipulated "control group". Also unlike an experiment we cannot control for confounding variables, things that change between the groups that we cannot make constant. Even so broad comparisons are possible. We know that modern democracies do not go to war against each other while countries run by dictators do. At our current stage of evolution it is clear what countries come closest to utopia. They are those that are democratically governed, have capitalist economies, tolerate openness and diversity of opinion, and value objectivity, reason, science and technology.

The closest we can come to an experimental manipulation is by investigating naturally occurring events. For example, crime rates were lower prior to the U. S. prohibition on alcohol, skyrocketed during that period and then came back down again after it was repealed. The lesson from this particular historical observation is that when a product is made illegal it drives up demand, creating incentives for criminal organizations to "set up shop". Without a system of laws in place to regulate competition, crime runs rampant as the different groups kill each other to control the black market.

Another way of determining which societies are better is to run simulations. The use of multi-agent systems (discussed in another essay) are ideal for this task as each "agent" can be programmed with a different set of values that determine its choices. Economic models now are quite simple. They explain things like consumer choice across a set of alternatives by having agents calculate which alternative produces maximum utility. More advanced computational capability will allow us to simulate the complex kinds of factors that actually go into choice and decision-making. We could for instance create simulations with competing values like cooperation vs. competition, or altruism vs. individualism then let the simulations with these different values systems run to see which produces better outcomes.

One of the problems with computer simulations of complex systems is that they lack long-term predictability. Thus, we may determine that one type of society emerges as near utopic early on in our simulations, but then discover later that it collapses into chaos and anarchy. There are also an incredible number of factors that determine the outcomes of nations. Computational models will be limited in the number of such factors they can include unless computing power is increased. Still, carefully constructed models can examine the addition or deletion of certain key variables and how they interact. This is a step toward a future discipline of computational culture.

Rather than simulate societies why not test them in real life? A group of libertarians since 2001 has been attempting to create a pro-liberty community of some 20,000 like-minded individuals in New Hampshire. Known as The Free State Movement, the goal is to concentrate their voting power and so leverage political action. Participants sign an agreement promising to move to the state within five years of the drive. As of late in 2014, there were over 16,000 people who signed the agreement. A far smaller number of them, only 1,654, have actually made the move.

The Seasteading Institute has proposed a more radical idea. This group would like to see communities living on floating cities. These pioneers could then experiment with new ideas for government and perhaps serve as an inspiration to change other ruling bodies around the world. Their stated first step is the Floating City Project, where a seastead will live in a host nation's protected territorial waters. Obviously this approach must overcome engineering challenges, produce economic incentives to entice migration and craft new governing policies without overt discord and chaos. A simpler solution might be to purchase an island from private or national owners and use that as the staging location instead.

The best way to create an ideal society is ideologically. That is, one needs to start with a philosophy of what it is to be human. If for instance, one determines that freedom is a primary human value, then a constitution and bill of rights can be set up around that value. The organization of the government and the types of laws that are created then become the means for allowing individuals to express that value. In this case we are being qualitative by establishing social parameters that protect our prioritized set of values.

CHAPTER 10. SPACE

PLANET HUNTING

Some say we are in a "golden age" of planetary discovery. In the past few years astronomers have confirmed the presence of many planets orbiting suns other than our own. These are called exoplanets. They have been able to do this more accurately through the development of new space-based telescopes. NASA's Kepler telescope is now in orbit around the sun. It is actively searching for Earth-sized planets among a group of stars that are 600 to 3,000 light-years away. It does this in part by detecting the drop in brightness that occurs when a planet passes between us and it's sun. The search thus far based on both terrestrial and space observations has been wildly successful. About 1,800 confirmed exoplanets with thousands more awaiting confirmation have been discovered as of late in 2014. Planets are so common it turns out that there is about one planet for every star in our galaxy on average.

There are a variety of different planets out there. As of November 2013, 674 have been classified as being Earth-size, while 1,076 are categorized as "Super Earth-size", being somewhat larger. The majority are bigger than this, equivalent in size to Neptune. Not all of these are habitable and most of them are gas giants incapable of supporting Earth-like life. Rocky planets tend to orbit near their stars. Examples of these in our own solar system are Mercury, Venus, Earth, and Mars. Gaseous planets like Jupiter are found orbiting farther away. Earth-like life would be most likely found on a rocky inner planet and not a gas giant.

There are an estimated 11 billion potentially habitable exoplanets in the Milky Way galaxy. One of the major goals of planet hunting is to find these planets, those that are about our size and at the right distance from their sun to support liquid water, the basis for organic life, as we know it. Two planets orbiting the red dwarf star Gliese 163 are within this habitable zone as are three other planets found in 2013 that may be covered in oceans thousands of kilometers deep. It has been estimated that an "alien Earth" may be as close to us as 21 light years away.

Much information can be gleaned about these exoplanets simply by telescopic observation. The atmospheric spectrum can tell us the type of gases that exist, whether oxygen is present and whether the combination of gases could have been generated by life. The presence of ozone, water, methane and carbon dioxide can also be detected. Oxygen and ozone on Earth come from living organisms, while bacteria can create methane. A more fine-tuned measure of the atmosphere can reveal the presence of nitrous oxide or artificially generated gases like freon that could be the calling card of an alien civilization. The presence of vegetation, mountains and seas would require bigger telescopes with higher resolution. Other features of exoplanets can be estimated, such as their daytime and nighttime surface temperatures, their surface composition, magnetic field, and whether they possess rings and moons.

The discovery of exoplanets has had a profound impact on astronomy. Even two decades ago it was unclear how many stars had planetary systems. Now we have developed a very good sense of the planetary population characteristics for our galaxy. Continued observation with bigger and better telescopes may confirm whether or not life exists on any of these worlds. In practical terms we may also determine which planets we could live on and which planets may be amenable to terraforming that could some day allow us to live there. Many of these worlds appear to be relatively close. We could send probes to them in a reasonable time if we are able to get the probes to travel at 10% or more of the speed of light. Colonization of these planets will take longer. Realistically speaking it will be several centuries before we can establish a colony on any of them. Development and settling of our own solar system is much easier and will take place first.

SPACE TRAVEL

If mankind is to survive we must spread out. Should any extinction level event occur to Earth it would be the "off-worlders" who survive and continue on. Humans far back into our history have been travelers. Several times during our early development in Africa we left the continent, passed into the Middle East, then travelled onward into Europe, Asia, the Pacific and the Americas. This dispersion too has contributed to our persistence. A war, disease or other disruption may devastate the local population but if distances are great enough it might have a negligible effect on other population centers.

Burt Rutan founded the company Scaled Composites to build craft that could take civilians willing to pay into near-space. His experimental craft won the Ansari X Prize for the first two flights by a reusable manned vehicle. It cost him $100 million dollars to develop it. The prize itself was only $10 million. But space fever seems to be catching on and there is money to be made in space tourism. There are 18 companies who are now in the act. The company Space Island Group has plans to build a private space station with rooms for rent. The cost to launch a payload has come down to about $70,000 for a single adult and will in all likelihood continue to come down making a trip to space a reality for the average person.

One prediction is that space travel will have a predictable pattern. It will start with satellites and planetary probes then progress to an Earth orbiting space station and moon base. After that we should see a Mars colony and asteroid mining and habitation. This would be followed by the colonization of Jupiter's moons and then the outer planets. Voyages beyond the solar system will originate with probes to nearby stars like Proxima Centauri and Barnard's Star and then human travel to stars with planets that are within a few light years reach from Earth. Travel within our own Milky Way Galaxy, adjacent galaxies and the remaining universe will follow.

There are many obstacles to interstellar travel. It takes a lot of energy to get a ship up to the high speeds required. A collision with any dust or gas going even a moderate fraction of the speed of light can cause serious damage to a ship or crew. Keeping crews alive for the long travel time is also difficult, as we currently do not have a viable method of suspended animation. Vacuum, radiation, weightlessness and micrometeoroids pose additional hazards. Interstellar probes are a much more efficient way of gathering information as they don't require life support infrastructure. One could manufacture many small probes and send them out in large numbers. This proposal has several merits. The probes are less likely to collide with objects because they are smaller. It also takes less energy to accelerate them.

The distances involved in even planetary travel are mind-boggling. At current spacecraft speeds it takes 8 months to reach Mars and 1.5 years to reach Jupiter. A journey to the edge of the solar system is over a decade. To give you a sense of scale, if the sun were the size of a beach ball the Earth is the size of a pea located 500 feet away! Jupiter would be the size of a tennis ball, located over 2,000 feet from the sun. The nearest star using this scale is 25,000 miles away. For extra-solar distances it makes more sense to use light years. Most bright stars in the sky are suns located within a few hundred light years from Earth. Our galaxy is 100,000 light years across and distances to near galaxies are measured on the order of millions of light years.

It is difficult to comprehend these vast distances even when given familiar examples. For instance if we were to shrink our solar system down to the size of a quarter the nearest stars are still on the order of several miles away! Our brains evolved in much smaller environments. We are used to thinking of distances in smaller units and tend to scale objects to our own size to see and travel. Adopting cosmic dimensions will take some getting used to.

If we can develop craft that travel a significant percentage of the speed of light then we can travel to local stars in our galaxy within the reasonable lifespan of a single generation. If this were not possible then we must learn to be patient. Human travel times in this case would need to be measured in terms of generations. Astronauts setting off now would die long before they reached their destination and it would be their grandchildren or great grandchildren that would be alive to experience the arrival.

Putting astronauts into suspended animation is an alternative. In any event, time travels faster for those on Earth relative to those speeding away. Those people we knew on Earth when we arrived would be long dead by the time we reached a distant star. The cost of transporting humans through space is expensive. In terms of information gathering it is a lot cheaper to use telescopes and unmanned probes.

Probes seem to be a much better option for exploration of both our own galaxy and adjacent stars. Electronic components keep getting smaller and it may be possible to pack a propulsion system, sensor suite, computer and communications equipment into a package the size of a baseball. It would take far less energy to propel this smaller mass up to fractions of light speed and because it is cheaper we could mass manufacture them. Imagine thousands of these probes sent out to all corners of our galaxy transmitting what they encounter along the way.

More of a technological struggle will be required to make von Neumann machines. These are probes that are self-replicating and named after the computer scientist John van Neumann who came up with the idea. These devices will be able to land on a planet and use its natural resources to manufacture copies of themselves. Once these copies are complete they can then be launched and sent on additional journeys to other planets where they could also land and build copies of themselves. Each machine as it encounters a new or significant discovery could communicate its findings to other probes in its vicinity, relaying the messages back to humans. With self-replicating probes we are not limited to the original number constructed, the total number would always be increasing exponentially. Going 10 percent of the speed of light it would take only a few million years to visit the entire galaxy.

Probes can serve many purposes. In addition to information gathering they could stay on certain planets and monitor them for signs of life or intelligent life. If we were really ambitious we could even have them terraform other worlds, seeding the conditions necessary for the emergence of biology. Given this may be within our future technological possibility one can ask why other civilizations have not already done this. Why have we not already encountered alien probes or life forms? This is the Fermi paradox referred to earlier. Three explanations for this paradox have been given. The first is that aliens may exist but make a choice not to communicate, what is known as the "zoo" hypothesis. It could be they are waiting for us to mature at which point they could greet or destroy us depending on how we turn out. This is the "berserker" hypothesis. Third, it may be they are so advanced they wouldn't even know how to recognize us, or would consider us inconsequential. This is the neophyte hypothesis.

So it seems we have three options for space exploration. The first are space telescopes that are best for long distance observation. This is the method of choice for understanding what is happening very far from us, at distances too great to send people or machines. These can tell us what is happening at the other end of our own galaxy or for phenomena outside our galaxy. For "intermediary" distances probes seem the best bet. These can effectively explore local space in and around our solar system including local stars that can be reached in a reasonable time. The last option of manned exploration is best for researching, mining, terraforming and inhabiting planets, comets and asteroids within our own solar system.

Part of what makes humankind unique is our desire to learn and understand. This understanding requires exploration, both the sedentary form in the laboratory studying a chemical reaction or the mobile kind tromping through a rainforest gathering specimens. There is only so much we can learn in the lab or the library, or for that matter from the telescope or probe. At some point to gain knowledge we must push out ourselves in physical form to see what's out there first-hand. Let us not become complacent or content with what we know. Nor should we be daunted by the challenges of space travel. We have begun this voyage already. Let's keep at it.

HOW TO GET THERE

Let's discuss space travel in practical terms. How will we actually get to the stars? Given the vast distances and unforgiving nature of space, is it even possible? The answer is yes. The technology to construct devices that will let us colonize our solar system and travel to others exists now or will become available in the near future. The first of these technologies is a space elevator that will drastically reduce the cost of transporting people and supplies off Earth. The second are designs for interstellar starships.

Rockets are expensive. They require tremendous amounts of fuel to launch even small payloads into Earth orbit. There is a better solution. An international team of scientists and investors has come up with an audacious plan – why not build an elevator that can take us into space? It may be possible to get started on such an adventure within a decade and NASA has offered prizes of up to $2 million dollars to those who can provide the necessary materials.

The scientist Bradley Edwards was the first to seriously consider how we might build a space elevator. He says there are three main components. The first is an elevator car. The second is a laser or other power source for the climber and the third is an elevator cable or ribbon. The cable would need to be made out of an incredibly strong material like carbon nanotubes and would originate at a mobile base in the ocean located at the equator. It would stretch to a space station in geostationary orbit located one hundred kilometers up into space. The Earth's own rotation would keep the cable taught and in position.

The elevator car would be an atmospherically contained chamber that people could stay comfortably in for the many hours the trip would take. The car would "crawl" its way along the cable using large robotic arms. A laser on the space-elevator platform would shoot a beam at a dish on the elevator to provide the necessary power. There are still obstacles to be surmounted on this project. We are not yet capable of manufacturing the carbon nanotubes or other materials for the cable. Also, the cost of such a venture would be enormous. As such it would almost certainly need to be an international project with funding from various participating nations.

Our trip through the solar system will take us first to the moon and then to Mars. Living on Mars is going to be difficult as we cannot breath the air or grow plants there as a food source. One solution is to terraform the planet. Ironically, we would want to create a greenhouse effect there, just exactly what we are trying to prevent here on Earth. This could be achieved by melting the Martian polar ice caps to release carbon dioxide into its atmosphere. The thicker atmosphere would retain more warmth and raise temperatures enough to allow for liquid water on the surface. This is a large undertaking and a long one. It could be several hundred years or even millenia before we could walk on the Martian surface breathing oxygen unassisted.

Michio Kaku outlines several types of starships that could take us to the stars. I will describe these here. The first uses a solar sail. In space, light can act as a "wind" pushing on a surface and drive it forward just as the normal movement of air propels a sailboat across the water. The Japan Aerospace Exploration Agency in 2010 successfully launched a spacecraft with a square shaped sail with a 60-foot diagonal length to travel to Venus. If we were to build a solar sail to power a craft on longer interstellar voyages it would need to be hundreds of feet across and gather momentum by orbiting the sun until it could slingshot away at a decent speed of say 10% of the speed of light.

Although it may seem crazy another way to propel a ship at fast speeds is to use nuclear bombs. The Orion project backed by such notable physicists as Freeman Dyson proposed just that. They had a design for a ship that would release hydrogen bombs from behind it and then detonate them at a distance so that the blast wave would propel the craft forward without causing damage. Conveniently, the project was a good way to reduce the number of nuclear weapons as it could eliminate as many as 2,000 of them on a single trip. The project was canceled when above ground nuclear blasts were forbidden by treaty in 1963.

An alternate proposal for a long distance spacecraft is a fusion ramjet engine. A scoop in the front of the craft would take in hydrogen gas that is abundant in space. It would then compress the gas using magnetic and electric fields releasing a large amount of energy that could be used as propulsion. Estimates show that a craft of sufficient size could accelerate to more than three quarters of the speed of light within a year. But there are problems. The scoop would need to enormous, about 160 kilometers across. Also, the ship as it moves through space creates a significant amount of drag that must be overcome.

Another high tech approach to creating a starship is to use antimatter. When antimatter and normal matter come into contact they annihilate each other producing a tremendous amount of energy that can be harnessed for propulsion. A major problem with this proposal is that antimatter is very difficult and costly to produce. Currently we are only capable of generating antimatter using colliders like those at Fermilab in the U.S. and the large hadron collider in Europe. These have been able to generate only small amounts, not nearly enough to drive a ship across the cosmos.

If we are to bring people into space we must overcome other hurdles as well. How do we sustain astronauts during these long transit times? The standard solution at least as portrayed in science fiction movies, is to put them in cryogenic suspension. They would be frozen in suspended animation and then brought back to life once they near the destination. This reduces the amount of atmosphere and food to be brought on board. As discussed in the longevity essay, cryogenics is at an early stage of development. To date no person has been successfully frozen for long periods of time and brought back to life.

An alternate solution is to construct a space ark. This would be a ship large enough to accommodate a reasonably sized crew that could live aboard it in comfort for many years if not centuries. The idea is to have multiple generations of families as flight crew. The downside to space arks is that they need to be large and completely self-contained and sustained living environments. Foodstuffs would need to be grown and harvested on board. Essentially it would mean recreating an entire ecosystem with plants and animals. All materials like urine and fecal waste would need to be recycled and reprocessed.

There are psychological issues here too. People may feel claustrophic or bored on long journeys and so entertainment would be necessary as a distraction. Virtual reality simulations might be one way in which the astronauts could immerse themselves so as to forget their current situation for a while. There is additionally the need to pick crews that would get along with each other. We would not want conflicts or fights to break out among the crew. All the people on board would need to be screened as being healthy and without any psychological disorders.

EXTRATERRESTRIAL LIFE AND INTELLIGENCE

A person's opinion on intelligent life elsewhere in the universe hinges upon several beliefs. The religious might consider it to be blasphemy. After all, God fashioned us in (his) own image, right? If there are extraterrestrials did God create them too? If so, then we are no longer unique. But if we look at the history of science we see a pattern where we are knocked further and further off our self-perceived "throne". Copernicus showed the Earth was not the center of the universe. Darwin showed we evolved like other animals. Now neuroscience and the field of artificial intelligence shows us that our minds are governed by natural laws just like the rest of the universe, making us hardly special at all.

Many people argue it is lonely and depressing should we be the only intelligent species in a universe so vast. They believe it is comforting to know there are others out there, especially others who are intelligent and with whom we can share thoughts and ideas. Whereas the first religious view advocated above seems to reflect arrogance and a "head stuck in the sand" mentality, the second more secular and scientific view reflects humility and humbleness when confronted with the mystery and wonder of the universe.

Other opinions stem on whether or not we think extraterrestrial life (ETL) will be hostile or welcoming. Just as is the case with AI, people seem to polarize into two camps. Some think these visitors will be aggressive, seeking to destroy us and harness Earth's resources for themselves. Others think they will be our saviors, providing us with advanced technologies that will solve all our problems and lead us to utopia. It is hard to say which of these views is more likely to be the case. Arguments can be made for either side. One thing for sure though is that if they visit us, they will be substantially more advanced technologically than we are simply by the act of being able to get here.

So is there intelligent life out there? If you run the numbers there most certainly seems to be. The astronomer Frank Drake devised a simple equation to estimate the possibility of extra terrestrial intelligence (ETI) in our galaxy based on certain pre-conditions. These include the number of stars in the galaxy, the number of those stars that have planets around them, the number of those planets that can harbor life, and so on. The actual probabilities for some of these conditions are not known and can only be guessed at so it is difficult to gain an accurate estimate. Of course the estimate falls even more on the side of ETI if we take the entire universe into consideration. There are many billions of additional galaxies in the known universe, so even exceedingly small values for these variables produce a positive result. The presence of life so far away from us though makes it difficult to detect or reach.

The Search for ETI (SETI) began as a search for radio signals only and so far has been inconclusive. But recent optical telescopes have had great success in detecting the presence of planets orbiting other stars. They have been able to deduce their presence by looking at how a star's light dims when planets pass in front of it. Far more planets have been found this way, and many closer to us, than was predicted by some. The amount of data that can be gleaned by some of these telescopes is remarkable and includes the type of atmosphere the planet has and even the presence of certain organic compounds. Future developments in such measuring instruments could yield even more detailed information that could answer additional questions concerning ETI.

If aliens came to Earth, how would they treat us? If these extraterrestrial beings are superior to us in intellect as they almost surely will be, does that give them the right to do as they see fit with us? After all, this seems to be the justification for the way we treat animals. We do pretty much whatever we want with animals because we consider ourselves superior to them. Should these alien visitors practice non-interference and only observe us until we have developed further? Should they gift us with wonderful new technology and elevate us to their level of understanding? Or would they be justified in destroying us and using our planet's resources? Let's hope Hollywood's answer to this isn't the right one.

FIRST CONTACT

How would humanity change if we made contact with an extraterrestrial civilization? The outcomes would vary depending upon the goals of this civilization, how far advanced it is technologically and the extent to which we could communicate effectively with them. Looking at the history of colonization by European powers in various locations around the world, the indigenous populations did not fare well. In most cases they were massacred, subject to slavery and converted to Christianity. At the very least such contact will profoundly change how we view ourselves in relation to the universe.

Distance is also an important factor. Contact with intelligent life located close to home in our own solar system would have a greater disruption than life located in distant star systems. Examples from human history show that the greater the distance between the two civilizations the less the contacted group feels a threat to itself and its culture. People close to a landing site would be affected more than those located farther away. People's beliefs also make a difference. In 2000, a poll of American and Chinese university students found that the more conservative the participants were the more they thought contact with an ETI would be harmful. Many of these students believed that the search for ETI would be futile as well as harmful.

If we make contact with a friendly ETI they could share advanced technology and a greater understanding of the physical world with us. Allen Tough suggests that if they had a history of warfare among themselves it would make them more likely to help rather than hurt us. He believes an ETI would most likely provide us with advice, consulting with world leaders and only act to prevent dangers such as a nuclear war with our consent. Contact with an advanced ETI, even if through radio waves and not physical presence, could produce a blow to our self-esteem and perceived uniqueness. Advanced ETI might understand this and mitigate contact, gradually exposing us to new knowledge and artifacts. Alternatively, they may seek to avoid all contact with us and let humanity develop further before initiating communication.

Scientists tend to agree that any ETI capable of interstellar travel would be much more technologically advanced than we are and thus easily capable of annihilating us. James Deardorff reasons however that there may be few aggressive species in the galaxy. Earth history shows that these types of civilizations don't last very long and that we are currently trending toward a more peaceful and democratic way of life. There are an abundance of natural resources in the galaxy and so conquering our planet in order to obtain resources is unlikely.

Astronomer Carl Sagan also adopted the peaceful scenario stating that any ETI that could come to Earth has probably transcended war in order to avoid destroying itself. Slavery may also be unlikely, as an advanced civilization would probably use robotic or automated labor. It could also be the case that civilizations that are technologically advanced would be ethically advanced and so not wish to harm us. There has been much debate about these issues in the astronomical community with opposing positions held by Seth Shostak and Ragbir Bhathal.

Whatever an ETIs level of technological development it is likely that humanity will experience shock and fear with first contact. The tribes in New Guinea, when first contacted by western nations, worshipped their technology and even created effigies of radio stations and airstrips in order to receive modern supplies and artifacts. This suggests that some people would form religions to worship the newcomers as Gods. The opposite could also come true, in which people of other faiths would see the visiting ETI as a threat or as demons sent to punish us. These groups might attempt to attack the newcomers.

Several studies have shown that most people would continue to adhere to their current faith given the arrival of an ETI. The Roman Catholic Church has even stated it would welcome their arrival. ETI does not challenge belief in a God and religions have in the past adapted to new scientific discoveries such as the news of the Earth orbiting the sun. Other more radical outcomes include the adoption of a new religion, one that includes all of humanity. This religion might stress the universality of human nature. Some people may adopt the philosophy or beliefs of the ETI itself.

The political reactions are varied. First contact could lead to a power struggle between rival groups with different agendas. There will inevitably be some debate as to which group would best represent humanity's interests. Perhaps a new council in the United Nations would be convened for such a task. If we were to reply to a radio message, there are many questions as to the format and content, particularly how we convey our history. If the distances are large with a reply taking many years, there will be ample time to discuss strategy and potential cultural impact.

How might we react if an advanced ETI decided to share their technology with us? Michaud suggests a two stage reaction. First, humanity would revere the information and its potential consequences, perhaps feeling a sense of technological inferiority complex. Second, we would gradually grow in confidence as we gained more knowledge and were able to apply it practically. We would need to be careful to study and use this knowledge prior to application in case it poses a threat. Possible threats could be instructions for how to create a powerful virus or other weapon.

CHAPTER 11. TIME

TIME AND INVENTION

Just as the space and distances of the universe are hard to imagine so are the time spans. The universe is estimated to be 13.7 billion years old while the Earth is 4.5 billion years of age. Living creatures only appeared on the scene 3.6 billion years ago. Humans began their evolution 2.5 million years ago and our current incarnation as anatomically modern humans wasn't until 200,000 years ago. Our entire existence is less than the blink of an eye in an unimaginably long stretch of time, a drop in a vast flowing temporal ocean.

To give you a sense of our ephemeralness consider time in terms of an average human lifespan. Let's set this to 30 years and consider it a single generation. To go back just to the beginning of recorded history roughly 6,000 years ago would then be 200 generations. That is a lot of great, great grandfathers. To go back to the first appearance of our first hominid ancestor would have been 83,000 generations.

These time spans should give us pause. Our race has accomplished much in a short period but perhaps the true mark of an advanced civilization is not so much what they have done but for how long they have done it. A great culture should overcome the threats to its existence, particularly the threat we pose to ourselves. It should spread out across the stars and colonize new planets, distributing itself across space as an insurance policy against destruction. It should learn to be patient, realizing that space travel is as much a battle against time as against distance.

As individuals we need to understand that to accomplish anything meaningful in our own lives takes a long time. It has been estimated that to become an expert it takes roughly 10,000 hours of practice. That is 3.5 years of working 8-hour days. If it is a hobby and one can only put in 2 hours a day it will take 13.8 years!

Different professions vary in their difficulty level. For neurosurgery it has been estimated at 42,240 hours. For yoga, it is only 700 hours. Of course if one were a genius these rates would be accelerated. Most of us only have time to acquire one career level skill. To obtain more than this means lots of extra work.

Much of technology is cumulative and builds upon what has already been discovered by science. The internal combustion engine could not have been built without an understanding of the physics of gas. Lights could not have been made without knowing the principles of electricity. Many people who are not researchers question basic science because they don't see how it is useful. What they don't understand is that *all* knowledge is useful. Understanding how the world works is the first step toward controlling and mastering it.

With regard to time we need to preserve not just knowledge but the things that have been built on that knowledge because they show how the process of invention works. An example of this is the Computer History Museum in Mountain View, California that chronicles the history of computing with many of the different models of computers up to contemporary times. Understanding how the engineering process proceeds and why certain approaches work while others do not can serve as a model for how to create other devices in the future. Humanity therefore needs a record of invention for the construction of devices, whether they are airplanes, elevators, or shoes. Inventors could access these records to use as a reference and for inspiration when developing something new.

By studying engineering in all fields it may even be possible to derive a general invention process that could apply to the design of anything. This might start with the desired goals of the device along with constraints on its operation. Then different design paths could be followed along with an evaluation of their success. This process could borrow from evolutionary principles but need not be identical to it.

LIFE EXTENSION

The science of life extension is the study of slowing or reversing the aging process. The goal is to extend life as long as possible, or forever, should that also be possible. There are a number of ways this can be achieved. The notion of life extension is a very popular idea and there is a lucrative market for it, with vitamins, herbs, hormone treatments and exercise programs all generating $50 billion alone just in the U. S.

The causes of aging are not well understood. According to the free radical theory aging is the result of atoms and molecules with unpaired electrons. These oxidizing agents are supposed to be a natural product of the cell's metabolic processes. They inflict damage on structures within the cells over time and it is this cumulative effect that is aging. The ostensible solution to this is to take antioxidants, substances like vitamin C or E that inhibit oxidation. Antioxidants such as these are also added to food to keep them from going bad.

Aging may not be inevitable. According to the Hayflick limit a cell in culture divides around 50 times before dying. Why is this? A structure called the telomere is supposedly to blame. A telomere is a repetitive piece of DNA that caps all chromosomes. They protect the end of the chromosome from deteriorating or from fusing with neighboring chromosomes. They get used up with each division and eventually there aren't enough left to allow the cell to reproduce. This suggests that each cell has a natural preprogrammed lifespan and that there may be a way to alter this programming.

There are some species that don't seem to age. A seabird called Leach's petrel has telomeres that lengthen as it gets older. Lobsters utilize an enzyme that prevents telomeres from degenerating. Charles Darwin brought back a tortoise named Harry from the Galapagos Islands to England when it was five years old. The turtle was eventually sent to Australia where it died in 2006 at the ripe old age of 176! Plants do even better than animals. An Aspen grove in Utah has been estimated to be 80,000 years old. A shrub in the Tasmanian rainforest is purported to be 43,600 years old and a creosote bush in the Mohave Desert is estimated as 11,000 years old. Understanding why these organisms don't age can give us clues to understand and prevent our own aging.

Longevity can also be measured at the species level. Ants hold the record. They have been around for 100 million years and are comprised of some 20,000 different species. There were some 17,000 species of trilobites that lived from the Cambrian to the Permian eras, a period of 250 million years. On average organisms made up of one cell like bacteria and plants last between 10 and 30 million years. It is a bit sobering to consider that in its history Earth has seen a half billion species of which only two percent are alive now.

Several methods available to us now have been tested to see if they extend lifespan. One of these, caloric restriction, has been found to work. Others, like growth hormone treatment have shown mixed results. Drugs that mimic caloric restriction like resveratrol and rapamycin have also failed to obtain consistent results, although they do have protective effects against cancers, cardiovascular and neurodegenerative diseases. Futurists have suggested alternative treatments such as nanotechnology and the cloning and replacement of body parts.

Cryonics involves freezing the body shortly after death in an attempt to bring it back to life later when medical science has supposedly advanced to the point where cures are available. No human or mammal so far has been successfully brought back to life after being frozen so it is still unclear whether cryonics works. Currently one can have their body frozen for $150,000. Your head alone is half this price (a sale?). In cryogenic techniques body temperature is lowered using liquid nitrogen to minus 321 degrees Fahrenheit. There are now about 200 people who are cryogenically frozen, mostly men and mostly entrepreneurs and techies. The two companies in the U.S. that provide this service are Alcor in Scottsdale, Arizona and the Cryonics Institute in Clinton Township, Michigan.

There are a number of ethical issues that arise from life extension and immortality. First, we must consider life quantity and life quality. If living longer means additional suffering then it may not be so welcome. Would you want to live to 150 if it meant 50 years of back pain or arthritis? If living longer involved this tradeoff, some individuals may choose not to do it. In fact, some people may not choose to do it at all even if this were not an issue. They might argue that life would become boring and meaningless if we knew we could live indefinitely. Why bother carving that sculpture or building that house if you could put it off until next century? Many of us are procrastinators as is. Immortality seems the ultimate excuse for doing nothing. It may reduce incentives or desires to be productive and creative.

Objections to the above are that some people are inherently curious. Many of us have interests or hobbies that we would like to pursue and with life extension we could focus on these full time. If one was sufficiently motivated and lived long enough they could taken on many different careers. Imagine how inspiring it would be for someone to be lawyer, medical doctor, scientist, and engineer and to excel in each of these fields with plenty of time to spare for retirement!

Overpopulation is another issue. If people live longer and continue to have the same number of children, the total number of people on Earth will go up. Longevity supporters deny this arguing that world population growth rates are decreasing and will continue to decrease. These so-called "immortalists" also argue that people who live longer will have fewer children or will space their children farther apart. It may be that growth rates will eventually level off or decrease due to worldwide improvements in nutrition, exercise and health care. This is the pattern we see in developing nations that modernize. Also, if we are able to stop the aging process it does not necessarily mean that we will be able to find a cure for existing diseases like cancer, Alzheimers and Parkinson's, meaning we could just as easily die a premature death by one of these as we do now. Certainly, famine, murder, wars and accidents could continue to take a toll should we not have a solution for them.

A related issue on aging is that the old will outnumber the young and we will be left with an aged society where the goals and values of the youth are ignored. This does not seem plausible. We actually live in an age now where there are an increasing number of elderly, those who belong to the baby boom generation. Yet our culture is youth dominated: most movies depict young attractive actors, most music is created for and listened to by younger adults. Many electronic products, cars and other goods are targeted for the youthful generation. This argument assumes we will continue to age, even with treatment, i.e., by merely slowing down the aging process. It may be the case that we could "freeze" the aging process at a youthful level, such as the low 20's and keep it there.

EXTINCTION

The universe is a dangerous place and humanity's survival is in no way guaranteed. There are a whole host of dangers that face us. Viruses like Ebola can spread quickly in a global economy. One infected person on an international flight could start a pandemic. Radioactive materials in dirty bombs and a traditional missile-driven nuclear holocaust are other still very real possibilities. In fact the world is so interconnected now that even a conventional terror attack (non-WMD) has far-reaching consequences. To illustrate, the September 11th 2001 attacks disrupted world economic markets for years. Dangers that could extinguish *homo sapiens* are referred to as existential risks. For a complete account see *Global Existential Risks* by Nick Bostrom.

The Earth has suffered several extinction level events that resulted in the near total elimination of all known species at the time. The best known among these was the asteroid strike that killed off the dinosaurs, but there are others like "snowball Earth" which was a severe planetary ice age that lasted an estimated 25 millions of years. There was also a period of super volcanism where giant volcanoes heated the planet and spewed poisonous ash into the atmosphere. Let us not also forget solar flares and gamma ray bursts from outer space that could fry us to a crisp. We are particularly vulnerable to radiation if the Earth's protective magnetic fields weaken or disappear which scientists admit is a possibility.

So what can we do about all of this? There are some efforts in place. NASA has a project devoted to mapping asteroids whose orbits could intersect the Earth. There are organizations like the Centers for Disease Control and Prevention (CDC), the United Nations (UN) and the International Atomic Energy Commission (IAEA) whose goals are to prevent various global threats, man-made or otherwise from occurring. Research institutes focusing on existential risk include Saving Humanity from Homo Sapiens (SHfHS), the Centre for the Study of Existential Risk and the Future of Humanity Institute.

This is not enough. As a species we need to wake up and realize that our collective survival is more important than our differences. If even one tenth of the effort we put into destroying ourselves could go toward saving ourselves, we would live in a much safer world. We need larger, better-funded organizations devoted to the scientific study of large-scale risks and their prevention. These need not be government-funded. They could be corporate and voluntary in nature. If Hollywood blockbusters are any indication, there are plenty of people scared about such events who will be willing to donate their time and money. In fact these movies could direct a certain percentage of their profits toward these organizations. Disaster sells.

Climate change and our response to it deserve some specific discussion. Even if climate change is not a threat it has focused attention on the relationship between humanity and our environment. Nobody actually favors pollution or prefers coal and oil to clean energy sources. The problem is really an economic one. The traditional energy market has an entrenched infrastructure and lobby organization. This makes start up costs for clean, renewable energies like solar, wind, hydro and geo exceedingly high. This is particularly true for third-world countries that lack the political will or resources to switch.

Free market environmental policies have yet to be given a fair try. Solutions like tradable quotas and property rights-based approaches avoid the downside of high cost across-the-board regulations. Rather than punish companies for polluting, it makes sense to reward them for producing clean energy. Species for profit plans have also had some success in saving extinct species. Instead of futile attempts to police poachers it makes more sense to open a zoo to allow visitors to see endangered species like tigers and elephants. Solutions like this legalize incentives and destroy crime-ridden black markets.

Another environmental problem as referred to earlier is overpopulation. There are some encouraging statistics though showing that as countries modernize their birth rates go down. Italy and Japan are now suffering a crisis of low birth rates. As countries like China and India catch up to the developed world we could see a leveling off or perhaps even a reduction in global population levels. Space travel and colonization will also reduce Earth's population in the long term and put less of a strain on our planet's resources. Asteroid mining and the utilization of "off world" goods in the centuries to come will constitute part of a new economy. Spacecraft production and the new technologies that accompany this off world migration could additionally serve to boost economic fortunes.

SAVING OURSELVES

It's a dangerous world out there, no doubt. Multiple times in its history the Earth has been subject to disasters that have wiped out large numbers of species. These mass extinction events have happened at least a dozen times. However, in each case life always came back. Life is in fact more resilient than we give it credit for. In one experiment microbes were left exposed outside the International Space Station for about a year and a half. Despite exposure to extreme cold and radiation, they survived. There is also now reason to believe that microbes can survive for even longer periods of time on asteroids and that life can be transferred from one planet to another by this method.

It is inevitable that another major extinction event will occur. It is just a matter of time. But there is hope that humanity might persist through one or more of these. We humans eat a lot of different things and can survive in any climate, which certainly helps. Annalee Newitz in her book *Adapt, Scatter, and Remember*, points out three strategies that might keep us alive. The first is to scatter. Humans even early on in our evolutionary history migrated out of Africa. She points also to the diaspora of the Jews around the world that has enabled them to survive despite constant persecution.

The second strategy is to adapt. Cyanobacteria are the ultimate adapters. They are able to generate their own energy from sunlight and can store food if needed. They are also capable of going into stasis for long periods if energy is not available and can tolerate extreme temperatures, hot or cold as well as acidic environments. Humans of course adapt themselves to different environments by cognitive rather than physical means.

Memory is the third skill. Creatures who survive and are able to pass on their survival strategies to offspring are also likely to persist. Gray whales are an example of this. They have one of the longest migratory routes of any animal along the western coast of the Americas. Younger whales follow the older ones and retain the route in memory. It was these skills that probably brought them back from the brink of extinction by being hunted by whalers. We humans also pass on survival strategies in the form of mythology, storytelling, literature and art.

Since large numbers of the human population are concentrated in cities, it makes sense to start with disaster preparedness in cities. San Francisco is in the vanguard of this effort in the U.S. The city has devoted billions of dollars, both private and governmental, into green spaces, energy efficient buildings, solar power, smart grid traffic systems and earthquake proof buildings. Implementing such procedures can be difficult though. There are zoning laws that restrict the kind of buildings that can be put up in different areas and there have been arguments over even using taxpayer dollars for disaster preparedness. The failure to update structures based on revised building codes has killed thousands of people who could have otherwise lived in countries like China and Haiti when earthquakes hit. Corruption and mismanagement played roles in these cases.

A new breed of science has emerged. Called disaster science, researchers in this field attempt to engineer the environment to prevent or reduce risk incurred during a calamity. The Energetic Materials Research and Testing Center in New Mexico measures the effects of explosions, gunfire and bombs on structures should these happen as the result of terrorism or natural causes. Texas A&M's "Disaster City" is an open-air facility where entire city blocks can be built and then destroyed. Recently they have been testing rescue robots that can fly or move about the rubble of a damaged structure looking for survivors. Other labs devoted to the study and prevention of disasters include Oregon State University's tsunami lab. They simulate the effects of tsunamis on model cities and coastlines. There is also a group at U.C. Berkeley examining how earthquakes damage buildings and ways of building or bracing them to withstand earthquake shocks.

Computers can also be used to save cities. IBM has developed a program called Smarter Cities that uses data, both historical and real-time, to predict and manage traffic patterns, evacuation routes, food delivery and energy distribution. The system has been used to successfully reduce traffic in Stockholm, Sweden. First responders like police, first aid, coast guard and government officials can use the data to determine how to direct help. Computer programs are especially helpful in epidemic modeling to predict the spread of infectious disease through a population. They can aid in efforts to quarantine and vaccinate. Since we are a global society the coordination of such efforts must be international.

Annalee next outlines a list of actions we could take to help survive the next apocalypse. Cities could construct underground habitations where large numbers of people can shelter from radiation or a meteor strike. There are psychological consequences to this however. Some people are claustrophobic and even those who aren't begin to suffer stress at being underground for an extended time period. Engineering solutions include making these areas more stimulating and varied as well as brighter and more spacious.

Another change we can make to cities is to make them greener by introducing parks and gardens both on the ground and on roofs. Plants sequester carbon and put out oxygen, helping to reduce climate change. Turning cities into "vertical farms" or what is called periurban agriculture, has also been suggested, mainly to reduce reliance on distant food sources. Beyond this architects are envisioning "biological cities" in which normally man-made structures are turned biological. Imagine curtains of algae that provide illumination by glowing at night, walls made of bioplastics instead of petroleum products and substances like BacillaFilla, a strain of bacteria engineered to produce glue and calcium. BacillaFilla when applied to cracks in walls seeps into the spaces, filling them up and repairing them. Bacteria can also be used to purify the air and serve as bioreactors for food and fuel.

It is possible to go to even greater lengths to protect ourselves. Geoengineering or terraforming involves altering a planet's weather or other major systems to bring about a desired change. One way to reduce global warming is to release sulfur aerosol exhaust. This makes clouds more reflective, bouncing back more sunlight and helping to cool temperatures. Another solar management technique involves the creation of thin disks of particles that will float in the atmosphere reflecting but not scattering light. These would be dispersed by weather balloons tethered to ocean-going ships.

There are risks to such methods. Critics have argued that they could erode the ozone layer and increase ocean acidification. Other suggestions along these lines include adding lime to seawater that would soak up carbon dioxide then allow it to sink to the bottom of the sea where it would remain. In a process called "enhanced weathering" rocks that absorb carbon dioxide could be broken into small pieces and exposed to the air. It is possible that some of these schemes might work but the general public may perceive them as "meddling with nature" and be too scared of the possible consequences to approve implementation.

As for foiling asteroids there are a number of possibilities. Blowing them up doesn't really help as they will simply fracture into smaller pieces that could cause just as much damage. NASA's program Spaceguard is an attempt to locate and track 90 percent of near-Earth objects (NEOs) larger than one kilometer. These can be asteroids, meteoroids or comets whose orbits put them in close proximity to Earth. The farther out we locate them the better as it becomes more difficult to deflect their course the closer they get to us. In the slow push technique spacecraft can fire lasers at an asteroid or comet. The ejected matter can serve like small rocket motors pushing the object in the opposite direction. It is also possible to push one of these objects by placing another object of smaller mass near it. The gravitational attraction of the two objects for each other would alter their orbit. Finally, there is the kinetic impact method that involves shooting a projectile into the object to move it. This has in fact already been done. NASAs Deep Impact mission used a probe to hit the comet Tempel 1 with a copper slug. Its orbit was altered by a small amount.

THIS IS THE END MY FRIEND

It is sad to say, but given our current understanding the universe will ultimately end. Chris Impey provides a detailed description of how this will unfold in his book *How it Ends. From you to the Universe.* To start off, our sun will not burn on indefinitely. It will get larger and brighter just as it has been doing for the last 4.5 billion years. Half a billion years from now this increase will move carbon dioxide from the Earth's atmosphere into the ocean. The results will not be good. Most plants will not be able to photosynthesize under these conditions.

Eventually the polar ice caps will melt raising sea levels. Following this water from the oceans will evaporate into space and the Earth's surface will become a barren desert. In 3.5 billion years from now the surface of our planet will be nothing but dried out rock. One billion years after that the sun will burn up its shell of hydrogen gas, bloating in size. Then it will become a red giant, 250 times larger than it is today and 2,700 times brighter. The Earth will become engulfed by the sun and eventually destroyed by it.

Our Milky Way too has a limited lifespan. Much of the gas necessary for star formation will eventually be used and there will be fewer new stars being generated. Eventually there will be only red dwarfs, neutron stars and black holes. In 10 trillion years even the red dwarfs will go out. Stars will no longer exist. These processes will be happening not just in our galaxy but in all galaxies across the known universe. The remnants of the Milky Way and other galaxies will not completely evaporate until 10 billion billion years from now, or 10^{19} billion years. The universe ultimately will lose its galaxies, any stellar remains being vacuumed up by black holes. Those black holes that existed at galactic cores or clusters will now merge with one another. In 10^{100} years protons will have decayed, all stars are dissipated and even black holes will have evaporated. All that is left will be neutrinos, electrons, and positrons, a particle soup left over from the universe's "glory days".

But don't lose heart. Our universe may only be one of a vast number of universes that are constantly being born and dying. The multiverse refers to the set of infinite or finite possible universes, including our own, that comprise all of existence. These universes include all of space, time, matter and energy. In the multiverse view there are multiple parallel universes that coexist together. The nature of these universes varies considerably depending upon one's discipline. Those that are most plausible come from the scientific fields of astronomy, cosmology and theoretical physics. For example, in one view our universe was born from the black hole of another universe and the black holes in our universe themselves give birth to other universes.

However, one must consider that these theories are just that: theories. We as of yet cannot prove the existence of any universe other than our own. How could we? By definition our universe consists of just those things we can observe, measure and devise scientific theories about. If we can't access another universe we cannot by definition test to see if it is there. Existence demands proof. If one cannot devise proof if one cannot prove existence. The cosmologist Paul Davies states that the multiverse question is really a philosophical argument, not a scientific one because it cannot be falsified. He believes that arguments about it are only harmful or quasi-scientific.

What is the psychological impact of a finite but very long duration universe? I think for most people it is depressing. Familiarity is comforting and the thought that all we know will eventually disappear seems tailor made to satisfy a roomful of anguished French philosophers. This notion goes beyond our own personal death. Most of us assume that after we die the world will go on. We derive some comfort from that, understanding that our children and grandchildren will continue to live out their lives, hopefully in happiness. But the notion of a terminal universe goes deeper than this. It states that at some point everything we know, everything we loved (or hated) will be gone, that all of existence, at least as we have experienced it, will vanish.

This is perhaps the reason why so many scientists believe in the multiverse. It satisfies that same need in us to believe in the afterlife, the "heaven" or "hell" that religious belief provides. Rather than our soul going off to heaven, our universe can instead go off and create other universes, reproducing itself in a never-ending process. This notion of infinity is more comforting than the finality of a single dying universe. Infinity is thus more than a parameter in a mathematical equation, It is a human construct, one that we want to believe because it makes us feel better.

There is another more optimistic take on this. It is finality that makes life worth living. Knowing we are mortal forces us to enjoy the here and now. It should make us savor every experience whether that be or a sip of that vintage Malbec wine or a lover's first kiss. It is endings that give meaning to life. If we were immortal or indestructible we may lose this sense of joy and wonder. Zorba the Greek had it right. There is more than enough in this world to satisfy us. We can never run out of things to learn, people to meet or places to go. The real tragedy is in not living life, in continually postponing our happiness thinking that we will be able to enjoy it later.

CHAPTER 12. PREPARING FOR THE FUTURE

TRANSHUMANISM

Transhumanism is a philosophical, cultural and political movement that believes advanced technologies can solve many of humanity's problems and improve humans beyond our current capabilities. They are technophiles who desire advanced human and machine intelligence, longer life and the augmentation of the human body. Its main governing body is The World Transhumanist Association, founded by Nick Bostrom and David Pearce in 1998. The major tenets can be found in the Transhumanist Declaration. These are sketched out below.

Humanity will be altered substantially by technology in the future. Some of the ways it will impact on us positively are through life extension, increases in intelligence both human and artificial, the capacity to alter our biological and psychological features, eliminate pain and suffering and migrate into outer space. We need to understand how to develop these techniques and assess their potential benefits and risks.

Transhumanists want us to embrace these developments and work on them in an open and tolerant environment. This is better, they argue than banning or prohibiting them. A case in point would be the banning of fetal tissue used to harvest stem cells. If we are to grow and progress as a species and have greater control over our own lives, then individuals must have the moral right to augment their mental, physical and reproductive capacities.

Of particular concern are the potential threats posed by rapid technological development. These should be anticipated, studied and planned for. Examples of such dangers could be a non-benevolent superintelligent AI, the spread of destructive nanorobots, or the propagation of a deadly, human engineered virus. Rational and public debates regarding the use of such technology will be necessary to determine the best course of action.

The transhumanists wish to support and help all sentient conscious entities whether they are regular people, genetically engineered humans or AI constructs. Their beliefs are largely in line with that of modern humanism but they do not support any particular political party, leader or ideology. Their connection to humanism is a belief that humans matter and that we can make things better through the promotion of reason, freedom, democracy and tolerance. A prime transhumanist virtue is autonomy. They want each individual to plan and choose what is best for him or herself. If some want to enhance, that is fine. If some desire not to, that is also fine, but the decisions of each should be respected.

Some of the technologies that will contribute toward human enhancement and happiness are AI, molecular nanotechnology, brain-computer interfaces and neuropharmacology. These can be used to control the biochemical processes in our bodies, eliminate disease and aging as well as increase our intelligence, emotional wellbeing and love for others. They could potentially enable us to have fine-tuned control over our desires, moods and mental states, to avoid being tired, hateful or irritated, to have an increased capacity for pleasure, artistic appreciation and to perhaps experience new states of consciousness. A future person whose possesses such features may no longer be recognized as human based on the standards we have today. He/She would become a posthuman.

There are certainly some among us who would regard the idea of a posthuman as abhorrent. They would object to implanting electronics in the brain or body for whatever reason, even if it were being used to treat a disorder and not to augment. Witness the Church of Scientology members who eschew many forms of modern medicine. If we view the human body as created by God, then to modify it is considered blasphemy by some. Others view the biological as natural and pure so combining it with technology is seen as denigrating and dirty. Just as we pollute the external world with industrial byproducts, some could see the introduction of such products into our bodies as a form of internal pollution.

FUTURES STUDIES

Futures studies is the study of what the future may be like. Just as history is the study of the past, Futures studies is the study of the future. While history relies on evidence, Futures studies relies on trend analysis and other methods in order to predict what might happen next. Futurism, as it is also sometimes called focuses on the "big picture". It attempts to understand broad aspects of culture such as political and social order, economics, religion, science and technology. It is thus interdisciplinary in nature but there have been recent proposals for how to unify its many disparate perspectives and methodologies.

The field distinguishes between possible, probable and preferable Futures, those that may be less like likely to occur, those that may occur and those that ought to or are desired to occur. Some futures studies researchers also examine "wildcards". These are low probability but high impact occurrences like an extinction level event. It also critiques different and/or competing views on contentious issue such as climate change. The concern is more with long- term predictions, those happening more than just several years ahead of the present.

Futures studies can trace its past to H. G. Wells, the British science fiction writer who was an early advocate, up to contemporary research centers at Universities in the U.S. and Europe. There are numerous organizations as well as books and journals that have been published on the subject. Some of its modern figures include Edward Cornish, Wendell Bell, Nick Bostrom, Peter Diamondis, George Friedman, Ben Goertzel, Stephen Hawking, Michio Kaku and Ray Kurzweil.

Futures studies assumes that there is more than one possible future. In other words there are many things that could happen and the question is how do we pick the right one? If there are alternate futures, how do we assign probability values to them to determine their likelihood of actualization? Researchers in this field use a variety of methods to answer such questions. One is called "backcasting" and involves identifying a desired future then traveling backward from it to determine what events in the present would need to occur to actualize it. Whereas forecasting is predictive and works from the now to the future, backcasting is proactive and works from the now to the possible future. Users of this technique don't just want to know what *will* happen, they want to *make* it happen.

In trend analysis a quantitative change in some variable over time, like computational power in Moore's Law, is projected forward and used to predict when certain events will occur. In reality there are often changes that can disrupt this projection by affecting the rate at which it occurs or by stopping and starting it, or stopping it altogether. Trends can come in all sizes, from "weak signals" to mega trends that extend over large periods of time.

Could we use mathematical models to predict future outcomes? Recent efforts suggest it may be possible. Bruce Bueno de Mesquita in 2011 developed a computational model to predict nuclear outcomes in Iran. His algorithm is based on game theory in which people make decisions based on self-interest and their understanding of what other people might do. Unlike most analyses that only take into account primary decision makers like presidents, his model factors in more individuals, those who would advise leaders and the groups that would in turn advise them. A CIA study found his model was right 90% of the time and could significantly outperform predictions by so-called experts.

Of course Futures studies is not without its difficulties. From dynamical systems and complexity theory we know that in natural systems small changes in initial conditions can lead to widely varying outcomes. This makes long-term predictability for systems like the stock market difficult. It is possible that advances in computer modeling may allow increasingly accurate long range forecasting in these areas.

The expression "Those who ignore history are condemned to repeat it" suggests that there are forces in history that create certain events and that awareness of these forces may allow us to prevent or promote the subsequent events. This is true in some cases but there are so many factors at work in history it becomes hard to generalize from one situation to another.

Despite these problems futures studies is gaining greater acceptance. It is used by companies to try and anticipate demand for products, by institutions that attempt to implement strategic plans and by governments as part of implementing public policy goals. It is also being taught around the world and some schools grant both undergraduate and graduate level degrees in this field.

YEAR MILLION

This book has focused on the immediate future based on trends and near term history. It is great fun though to speculate on what the far future might have in store. In *Year Million*, Damien Broderick collects essays by fourteen leading scientists and science writers asking them to hypothesize what the future of humanity will be like in one million years in the future. This is obviously so far ahead that any sort of accurate prediction is meaningless but it is interesting to see what sorts of ideas were proposed. Some of these will be sketched out here, especially those that are new and that have not already been covered.

One of the first questions that come up concerning the future is whether or not we will be around to see it. In other words, will humans persist for another million years? Richard Gott III, an astrophysicist at Princeton University uses a statistical procedure to estimate humanity's longevity. It is based on how long we have already been around and a 95% confidence interval that we will continue to be around. Given our existence now for 200,000 years he estimates we can be 95% sure that we will be around for at least another 5,100 years but that we will disappear within 7.8 million years. This places us within the expected lifespan of other hominid species (1.6 million years) and to species of mammals in general (2 million years). So by this estimate there is a good chance that we will still be around a million years from now.

Steven B. Harris and others have proposed that humanity will develop into a merged group mind. This type of composite mind is more intelligent than any single person and in fact is already in existence in the form of cooperative groups of people sharing information over the Internet. A bank, stock market, military organization or even a supermarket are all run by a group of people who work and communicate back and forth with one another to achieve a whole that could not be done by any one of its members in isolation. Electronic communication like email and phone conversation facilitate information flow between members and can allow for large numbers of individuals to participate. We can think of these linked groups of biological brains and technological computers as part of a superintelligence.

In the far future the speed with which members of these networks can communicate with one another will increase. This rate of information transfer will allow for more sophisticated abilities. A neuron transmitting and receiving messages with thousands of its neighbors forms the basis for human cognition. Similarly a person sending and getting messages from many other people in real time can form the basis of group cognition. This type of social computation may not even require a central organizer to coordinate activity, nor may it require individual members of the computation to be aware of what the purpose or goal of the computation is. Tasks could get executed through self-organizing emergent properties and without individual conscious awareness of the participating components, as is the case with human brains and other biological networks.

One theme a number of authors keep touching on is how radically different our bodies will be in the future. If we assume genetic engineering can be perfected then we may be at liberty to make extreme adaptations of our bodies, with wings enabling us to fly, gills enabling us to swim, squat strong bodies to adapt to large high gravity worlds, etc. Our definition of human will be stretched to the limit, writes Wil McCarthy. He envisions a future of androids, humanoids, hive minds and living vehicles. In fact, we may be become the aliens we expect to find!

Catharine Asaro believes we will see communities of like-minded and like-bodied individuals deciding to colonize a planet and create their own society based on shared norms, like the Amish people in the U.S. do now. The great distances and differences between settled habitats in space may see divergent evolution take place, accentuating differences between human groups both physically but also culturally in terms of religious beliefs, political views and sexual orientation. Despite these radical changes, Catharine believes there will still be a "family" unit that persists because people will want the companionship of a mate, a means to raise children, or pooled economic resources. She thinks the desire to love will remain as well. Although it may be expressed differently we will maintain love, friendship, and the need to form relationships. Given that the power to create these traits will be in our grasp, the decision to alter ourselves must be democratic and open to debate.

Robert Bradbury takes megaengineering to its absolute limit. He argues that humanity will ultimately be capable of building Matrioshka brains (Mbrains). Named after the Russian dolls that are nested one inside another an Mbrain will be a series of increasingly larger shell like structures around the sun. If these shells were completely enclosed they would be unstable, so he envisions a collection of co-orbiting solar sail-like structures. Each concentric level of shells would utilize energy from the sun. The shell behind it would harness the waste heat generated from the adjacent interior shell. Every shell would be made up of a substance dubbed computronium, a substrate devoted to computation. Bradbury believes we can upload our minds into the computronium and exist as a shared group mind there, inhabiting virtual realities alongside artificial intelligences.

The creation of an Mbrain requires the dismantling of asteroids, comets and planets to provide raw materials. Planetary break up could occur by using immense thin mirrors to redirect the sun's energy. Valuable minerals like aluminum, silver, gold, iron and related elements could be extracted this way. Of course the destruction of a planet may be considered a criminal act, just as terraforming it might.

Would we be willing to go to such extremes? Bradbury thinks we will, given the result is a computer the size of a solar system. A computer like this could perform astounding feats, emulating the entire history of human thought in microseconds and taking mere seconds to run thousands of thousand-year scenarios. The computronium could hold vast sums of knowledge like galactic and civilization history and detailed genomic plans for all known organisms. An Mbrain could devote itself to solving impossible tasks like how to reverse the expansion and decay of the universe or how to open up gateways to other universes.

Mbrains could ultimately duplicate themselves and spread to different stars. One can even imagine a galactic scale Mbrain using energy derived from a supermassive black hole at the galactic center. This could be linked to solar Mbrains around each star, uniting in a vast galactic computing network. That in turn could be linked to other galactic scale Mbrains uniting the entire universe as a single computing entity.

Whew, we're getting a bit ahead of ourselves here! Collective minds inside an MBrain would only be possible if mind uploads/whole brain emulation were possible. Supraluminal communication between Mbrains would also be necessary to coordinate any collective activity amongst them and so far as we know this isn't possible either. Note that many futurists predict the universe will progress this way, by converting energy and matter into computation, essentially turning the universe into a single brain or conscious entity, perhaps the mind of God if framed religiously. Kurzweil's last stage specifies this evolution as do those advocating the Omega Point and Aleph state.

CONCLUSION. HOW TO SAVE THE WORLD IN 65 EASY STEPS.

We have covered a lot of topics in this book and it is time to summarize some of what we have learned. In this section we will run through the essays covered and extract from them essential lessons. They are presented here under their relevant chapter headings and in the order presented throughout the book. Some of these summaries can be considered as recommendations for a better future, or suggestions for how to "save the world".

Chapter 1. Knowledge and Knowing

1. We may be living in a simulation but until this is revealed there is no reason to be alarmed.

2. There are a limited number of ways by which we can determine the truth. These include reason, logic and the scientific method. Mysticism, emotionalism and intuition are appealing but invalid epistemological methods.

3. It is important to educate the public on the procedures mentioned above. Critical thinking, mathematics and computer programming are also crucial skills as is the ability to be skeptical. Don't believe everything you hear or see. Always challenge assumptions and think for yourself.

4. Games have been incorporated successfully into educational programs. They may also be a new method for solving widespread social problems. We should give them a try.

5. Much valuable information is lost during the transition between old and new technologies. There should be some attempt to capture this information and preserve it for historical analysis.

Chapter 2. New Science

1. The field of quantum mechanics although bizarre has been shown to hold up under empirical testing and is mathematically consistent. We must accept it but try to integrate these findings with larger-scale physics. The universe is strange and our everyday experience and common sense can't always help.

2. Many systems in nature are self-organizing. They generate order without any external planner. These systems also contain different levels. The levels seem to emerge spontaneously. We need to understand how emergence and self-organization work and then use them to build and design our own systems.

3. Traditional science should be supplemented by the dynamical systems approach that focuses on nonlinear phenomena, interdependence and a holistic perspective.

4. Complex systems in nature cannot be predicted in the long run even though they may be described in a deterministic way. This reflects a fundamental misunderstanding of the way they work. Improvements in computational modeling may help.

5. Network science is another recent development in the sciences that stresses the study of systems independent of their substrates. There are architectural and functional similarities among networks of all sorts, whether they are the Internet or a brain. The study of one network can give us insights into how others function.

Chapter 3. New Technology

1. Technological progress is happening at a rapid exponential pace. We will probably continue to see tremendous increases in the power and sophistication of technology. It can be used for good or ill however and we must be careful to guard against its dangers.

2. The world is a much better place than it used to be thanks in part to technology. We should provide incentives for innovative technological solutions to world problems like water, food, housing, energy, health and freedom.

3. Individual genome mapping will become cheap and widespread as a diagnostic for detecting and preventing disease. Techniques like gene therapy, stem cells, organ replacements and nanobots show great promise for treating disorders.

4. In all likelihood the difference between biology and technology will narrow as we move into the future. Technology will be imbedded in our bodies or even ingested. The machines won't want or be able to wipe us out because "they will be us and we will be them".

5. Future technological developments will scale both to the very small and the very large, being nanotechnology and megaengineering, respectively. Both have great promise. Nanotechnology could bring about a revolution in medicine. Megaengineering could solve our transportation and energy problems.

Chapter 4. Intelligence

1. Rather than making us dumb technology may be making us smarter, at least as measured by gains in I.Q. scores in the general population over time.

2. Human intelligence can be characterized as generating hypotheticals, or "What if?" situations and using them to act successfully in the world. AI programs based on this method may become smarter or perhaps self-conscious.

3. AI programs are better at performing specific tasks than people, but human intelligence is more general and successful in a wider

range of situations. One of the goals of AI is to create an artificial general intelligence. We should work toward this goal.

4. The singularity is the point at which an advanced AI can improve itself without human assistance. Once this happens, it may quickly outstrip us in terms of intelligence. It is unlikely that the singularity will pose a threat to humanity if technology and biology continue to merge.

5. It is difficult to predict how a superintelligence might act but it is probably prudent for us to build in safeguards. These could include top-level goals of benevolence and the protection of human values.

6. Autonomous AI systems could perform malicious acts as an unintentional consequence of pursuing ethical or neutral goals. To prevent this we can utilize a staged scaffolding technique where we monitor and evaluate AI performance.

Chapter 5. New Beings and New Beginnings

1. Life may be non-corporeal, given the emergence and self-organization of complex biological features in artificial life simulations. Life and intelligence may also arise from multi-agent systems.

2. The principles of evolution can be applied to both software and hardware-based agents. Robots can be evolved to perform tasks using the principles of natural selection.

3. An artificial person is an agent that for all functional purposes is indistinguishable from a normal human being. Unlike an android, an artificial person need not look like a person. Artificial people would need to be afforded rights and treated the same as present day humans.

4. Advanced computer simulations could be used to model social behavior. The results could help diagnose and determine solutions to social problems. For example we may one day use them to judge what the best governments and economies are, but only if they are tested and validated against empirical data.

5. Governing robot and robot-human societies poses the same sorts of problems encountered in governing human-only societies. We may need to resort to behaviorist, rule-based or virtue approaches.

6. Cyborgs are a combination of "man and machine" involving both biological and technological elements. We will see more of these in the future. They will manifest in the form of neural prosthetics, nanorobots and other as yet unknown technologies. It seems certain that humans will evolve into bio-tech hybrids.

7. In the next century we will be able to modify some human characteristics genetically. This raises a number of ethical issues. There will be consensus about eliminating disease but not on what abilities should be enhanced or who will have access to such procedures.

Chapter 6. Toward a Future Psychology

1. Consciousness is the only thing that really matters in the universe, as it is only conscious entities that can experience meaning. We may need a new system of ethics to determine which rights apply to which types of new conscious entities.

2. If consciousness is based on hardware-dependent neural activity (wetware) then it will not be possible to perform a "mind upload", i.e., a transfer of human consciousness into a machine. The hardware/software distinction applies to computers not brains. Positing the existence of human "software" can be considered a type of dualism.

3. The concept of self is best characterized by a consistent way of thinking or acting that can be described using traits. People will interact with others electronically using avatars more frequently in the future. Digital avatars as computer programs will persist after our death and can continue to interact with the living.

4. Free will and determinism are best considered as points on a continuum and not as an either-or distinction. The more decision-making is centralized inside an agent, the more control it has over its behavior and the freer it is to act. Consequently humans have more free will than animals.

5. Humans are both rational and emotional. We need to learn to control our emotions; otherwise we are no different than the animals.

6. As technology evolves entertainment will become more sophisticated and specialized. We will be able to immerse ourselves in compelling virtual worlds or get high using precisely controlled recreational drugs. If this is the case there is a danger that society could become more hedonistic and decadent, losing sight of important values.

7. Sex with and love for robots will be more common as robots become more human-like in appearance. This will ultimately be accepted by society given the historical trend toward sexual liberalization.

Chapter 7. A Higher Calling

1. Humanity needs to objectively determine what our higher-order values should be (along with the virtues needed to achieve them). These could include freedom, knowledge, productivity and peace. In an ideal world, people should strive to achieve these rather than satisfy basic motivational drives.

2. It is better to have a philosophy than to believe in a religion. Religion is based on faith rather than reason. Philosophies are best used to tell us how to live a good life. Experiencing the beauty, vastness and intricacy of the natural world can satisfy our need for spirituality.

3. In the next stage of human evolution we will transcend our current limitations. It is not clear how this might happen but it could be by integrating our brains with superintelligent computers. This could foreseeably allow us to access all of the world's stored knowledge and to interact with other similarly connected minds. The singularity may thus be a human experience rather than one experienced by a machine separate from humanity.

4. The anthropic principle, Omega Point and other theories that posit a multiverse or conscious universe are purely speculative and theoretical. There is no evidence to support these claims.

Chapter 8. Social Systems

1. Genetic engineering may allow us to eliminate violence and aggressive behavior. While doing this we need to be careful not to eliminate other related personality traits like ambition and competitiveness.

2. Democracy is the best but not necessarily ideal form of government. There has been a trend toward greater democracy globally. Technology can enable democracy by allowing greater voter participation in the political process. Mass communication can help groups rally for demonstrations against repressive governments.

3. The need for physical money may disappear in the future to be replaced with electronic exchanges. Private competing currencies

like Bitcoin can potentially prevent devaluation and spur economic growth but there are security concerns that must be addressed.

4. Technology will provide increased product information to consumers allowing them to make more informed choices. Providers will be able to track customer preferences based on search histories and tailor advertising of specific products.

5. The ideal energy is clean, cheap and renewable. Our current use of gasoline and coal is contributing to climate change and is producing adverse environmental consequences. Solar, wind, hydro and geothermal are all better options. We should incentivize and switch to them as soon as possible.

Chapter 9. Civilization

1. Humankind has been seen as passing through a series of stages based on information and energy usage. Advances to human civilization will thus be measured by how efficiently and on what scale we can utilize these resources.

2. Dynamic societies that are open to creativity and entrepreneurship will see the greatest amount of innovation and wealth creation. Static societies that are too fearful of change and risk will stifle economic growth. Some amount of risk is necessary to drive society forward.

3. Technology enables creativity and makes it easier for individuals and companies to create novel art, ideas, and products.

4. Happiness is the main purpose of life. Freedom and opportunity are necessary for individuals to pursue happiness. A society is civilized to the degree that it allows individuals to do this.

5. It is difficult for a repressive government or empire to enforce its will on others across large distances. The difficulties involved in

projecting force between planets in space may serve as an (at least initial) effective buffer against tyranny.

6. Secession if done democratically and bilaterally ought to be allowed, as should the desire to rejoin a larger nation or confederation of states. Autonomy at the local level supercedes patriotic allegiance. All nations and states should have the right of freedom of entry and of exit.

7. A perfect utopia is probably not possible, but it is viable for cultures to maximize certain values like freedom.

Chapter 10. Time

1. The universe and our planet are very old and we have only been around for a tiny fraction of this time. We need to learn patience both as individuals and as societies when it comes to pursuing our achievements.

2. Life extension is possible using caloric restriction and possibly other methods but for how long and under what quality of life is unknown. Although it may be possible to live for a very long time, it is probably not the case that we can become immortal.

3. Overpopulation of Earth is unlikely. Birth rates go down as nations modernize. Technology allows for more efficient utilization of food and other resources. Migration off planet will reduce the planet's population but probably won't happen for at least a century or more.

4. There are many existential risks that face humanity, both from the environment and from our selves. There needs to be greater international cooperation on determining the relative danger of these threats and how to cope with them.

5. Dispersing, adapting and remembering responses to previous disasters can contribute to our continued survival. So can research in the fields of disaster science, disaster simulation, and disaster management. We need to continue to monitor potentially hazardous objects that could collide with the Earth.

6. Our solar system and the universe as a whole will come to an end in a far distant future. Rather than a cause for sorrow this should help us better appreciate the joys of life.

Chapter 11. Space

1. The most economical way to learn about other planets is through space telescopes that can provide us with a wealth of information including the size, location, atmospheric makeup and presence of life on other worlds.

2. The barriers to space travel are formidable but not insurmountable. We are a species of explorers and adventurers. This is how we have learned and understood the world around us. Let's continue the voyage.

3. We currently possess or are likely to possess the technology that will allow us to colonize the solar system and travel to other stars. The effort will be expensive and challenging but necessary if we are to survive and progress as a species.

4. Although there is no physical evidence, there are good arguments made in favor of extraterrestrial intelligence.

5. First contact scenarios vary widely. Some people would welcome ETI; others would fear them even if their intentions were benevolent. Any type of contact with an ETI would transform our view of ourselves as being somehow special or exceptional.

Chapter 12. Preparing for the Future

1. The transhumanist movement advocates the use of technology to improve the human condition. The stated goals of the movement are to extend life, increase intelligence, improve our biological and psychological traits, eliminate pain and suffering and migrate into outer space. These are laudable goals but we need a larger public discourse on how they should be implemented.

2. Reseachers in the futures studies movement attempt to predict the future and bring about positive future outcomes. They utilize a variety of methodologies. There are difficulties in attempting to do this and many predictions may be only partly true, but the movement is now an established academic discipline.

3. It is difficult to be accurate predicting what will happen in the far future but several leading writers and scientists have suggested what they think the world will be like in one million years. Group minds, extreme genetic body modification, divergent human evolution and the construction of enormous computers have been proposed.

So there you have it. These predictions and recommendations are based on actual past and current events and anticipate mostly the near future to insure accuracy. It is entirely possible that only some of these will come to pass. Regardless of this I believe the advice is sound. A reader may have noticed that ideologically I am an objectivist and libertarian. Some of this influence is evident in the essays. However, I make no attempt to defend these positions, as those are done more effectively elsewhere.

We humans are a mixed bag. We are capable of great kindness and love but also of profound hate and destruction. We can erect magnificent temples and buildings but also bomb them into rubble. We can write fantastic works of fiction and philosophy but then throw them into a bonfire to burn. We understand a lot about the universe around us but still have much to learn. Statistics show the world has become a better place in recent decades. This is an encouraging sign that we are on the right track. Let's make the correct decisions and continue the trend.

APPENDICES

Suggested Readings. The following readings are for the most part non-technical introductions and do not require prior knowledge of the field.

Anderson, T. L. & Leal, D. R. (1991). *Free market environmentalism.* San Francisco, CA: Pacific Research Institute for Public Policy.

Bell, W. (2010). *Foundations of futures studies: Human science for a new era: History, purposes, knowledge.* Transaction Publishers.

Blackmore, S. (2013). *Consciousness. An Introduction. (2nd Ed.)* London, UK: Routledge.

Bostrom, N. (2014). *Superintelligence: Paths, dangers, strategies.* Oxford, UK: Oxford University Press.

Bostrom, N., Cirkovic, M. M & Rees, M. J. (2008). *Global catastrophic risks.* Oxford, UK: Oxford University Press.

Brighton, H. & Selina, H. (2003). *Introducing artificial intelligence.* Cambridge UK: Icon Books.

Brockman, J. (2010). *This will change everything. Ideas that will shape the future.* New York: Harper Perennial.

Broderick, D. (2008). *Year million. Science at the far edge of knowledge.* New York: Atlas & Co.

Christian, D. (2004). *Maps of time. An introduction to big history.* Los Angeles: University of California Press.

Chua, A. (2007). *Day of empire. How hyperpowers rise to global dominance – and why they fail.* New York: Doubleday Press.

Cornish, E. (2005). *Futuring: The exploration of the future.* World Future Society.

de Grey, A. (2007). *Ending aging: The rejuvenation breakthroughs that could reverse human aging in our lifetime.* St. Martin's Press.

Drexler, K. E. (2013). *Radical abundance: How a revolution in nanotechnology will change civilization.* New York: Perseus Books.

Friedenberg, J. (2008). *Artificial psychology. The quest for what it means to be human.* New York: Psychology Press.

Friedenberg, J. (2009). *Dynamical psychology. Complexity, self-organization and mind.* Emergent Publications.

Hammersley, B. (2013). *Approaching the future. 64 things you need to know now for then.* Berkeley, CA: Soft Skull Press.

Hofstadter, D. (2007). *I am a strange loop.* New York: Basic Books.

Hudgins, E. L. (2002). *Space. The free market frontier.* Washington, D.C.: Cato Institute.

Ichbiah, D. (2005). *Robots. From science fiction to technological revolution.* New York: Abrams, Inc. Publishers.

Impey, C. (2010). *How it ends.* From you to the universe. Norton: New York.

Istvan, Z. (2013). *The transhumanist wager.* Futurity Imagine Media LLC.

Johnston, J. (2008). *The allure of machinic life. Cybernetics, artificial life and the new AI*. Cambridge, MA: MIT Press.

Kaku, M. (2011). *Physics of the future*. How Science will shape human destiny and our daily loves by the year 2100. New York: Anchor Books.

Kaufman, M. (2011). *First contact: Scientific breakthroughs in the hunt for life beyond Earth*. New York: Simon & Schuster.

Kurzweil, R. (2006). *The singularity is near. When humans transcend biology*. New York: Penguin.

Kurzweil, R. & Grossman, T. (2009). *Transcend: Nine steps to living well forever*. Rodale Press.

Levy, D. (2007). *Love + sex with robots*. New York: Harper.

Levy, D. A. (2010). *Tools of critical thinking. Metathoughts for psychology*. Long Grove, Il: Waveland Press.

Maslow, A. H. (2013). *A theory of human motivation*. Start Publishing.

Newitz, A. (2013). *Scatter, adapt, and remember*. New York: Doubleday.

Paul, R. (2007). *A foreign policy of freedom. 'Peace, commerce, and honest friendship'*. Lake Jackson, TX: The Foundation for Rational Economics and Education, Inc.

Postrel, V. (2011). *The future and its enemies. The growing conflict over creativity, enterprise and progress*. New York: Touchstone.

Schmidt, E. & Cohen, J. (2014). *The new digital age. Transforming nations, businesses, and our lives.* New York: Vintage.

Schneider, S. (2009). *Science fiction and philosophy. From time travel to superintelligence.* West Sussex, UK: Wiley-Blackwell.

Shanks, P. (2005). *Human genetic engineering: A guide for activists, skeptics, and the very perplexed.* New York: Nation Books.

Wallach, W. & Allen, C. (2009). *Moral machines. Teaching robots right from wrong.* Oxford, UK: Oxford University Press.

Wilson, E. O. (1999). *Consilience: The unity of knowledge.* New York: Random House.

Web Resources. These are organized topically in alphabetical order.

Artificial Intelligence (AGI)

http://wiki.opencog.org/w/The_Open_Cognition_Project
http://wp.novamente.net/
http://www.agi-society.org/

Artificial Intelligence (Friendly AI)

http://www.goertzel.org/dynapsyc/2004/PositiveTranscension.htm
http://selfawaresystems.com/
http://friendly-ai.com/

Artificial Intelligence (General Issues)

http://www.aaai.org/home.html
http://www.corticaldb.com/
http://www.lib.hust.edu.cn/xueke/xtgc/xh/aaai.htm

Complex Systems

http://necsi.edu/
http://santafe.edu/
http://www.complexity.org.uk/complexityscience/

Existential Risk

http://shfhs.org/
http://cser.org/
http://www.existential-risk.org/

Futures Studies

http://www.wfs.org/futurist
http://www.fhi.ox.ac.uk/
http://www.iftf.org/home/
http://thefutureoflife.org/home

Life Extension

http://www.longecity.org/forum/page/index.html
http://mfoundation.org/
http://www.genescient.com/
http://www.lifestarinstitute.org/
http://www.sens.org/

People

http://www.nickbostrom.com/
http://wp.goertzel.org/
http://mkaku.org/
http://www.kurzweilai.net
http://stevenpinker.com/
http://vpostrel.com/
http://www.optimal.org/peter/peter.htm
http://yudkowsky.net/

Presentation and Debate

http://edge.org/
http://www.ted.com/
https://www.sciencenews.org/

Rationality

http://www.overcomingbias.com/
http://lesswrong.com/
http://www.criticalthinking.org/

The Singularity

http://singularityhub.com/
http://singularityu.org/
http://intelligence.org/

Social Systems and Organizations

http://www.seasteading.org/about/visionstrategy/
http://www.cato.org/
https://www.theihs.org/
http://www.atlassociety.org/

Technology

http://www.ieet.org/
http://www.seriouswonder.com/
http://www.nano.gov/nanotech-101/what/definition

Transcendence

http://brainmeta.com/index.php?p=consciousness-singularity
http://www.aleph.se/Trans/Global/Omega/
http://www.ibiblio.org/jstrout/uploading/

Transhumanism

http://www.extropy.org/
http://humanityplus.org/
http://www.aleph.se/Trans/Intro/index-2.html

Science fiction media. The following list of science fiction media comes from the book Science Fiction and Philosophy. From Time Travel to Superintelligence. Edited by Susan Schneider. They are organized topically.

Computer Simulation and Virtual Reality

The Matrix Trilogy
Permutation City
The 13th Floor
Vanilla Sky
Total Recall
Animatrix

Free Will and Personhood

Software
Star Trek, The Next Generation
Second Chances
Mindscan
Minority Report

Kinds of Minds

2001: A Space Odyssey
Blade Runner
Artificial Intelligence (AI)
Frankenstein
Terminator movie series
I, Robot

Ethics and Social Issues

Brave New World
Gattaca
White Plague

Space and Time

Twelve Monkeys

Slaughterhouse Five
The Time Machine
Back to the Future
Flatland: A Romance in Many Dimensions

www.ingramcontent.com/pod-product-compliance
Lightning Source LLC
Chambersburg PA
CBHW060557200326
41521CB00007B/600